To
Thomas William Webb
who has given us all so much pleasure
and to
Peter Hingley
who in 1997 decided that it was high time somebody did
'Something Definite About Webb'

The Stargazer of Hardwicke

Revd Thomas William and Mrs Henrietta Webb. (From the photograph pasted into the front of Webb's fifth observing notebook.)

The Stargazer of Hardwicke:

The Life and Work of Thomas William Webb

edited by

Janet and Mark Robinson

GRACEWING

First published in 2006

Gracewing
2 Southern Avenue
Leominster
Herefordshire HR6 0QF

ISBN 0 85244 666 7

Typesetting by
Action Publishing Technology Ltd., Gloucester, GL1 5SR

Printed in England by
Biddles Ltd., King's Lynn PE30 4LS

Contents

	List of illustrations	ix
	Contributors	xi
	Foreword: *Sir Patrick Moore*	xv
	Editors' preface	xvii
	Acknowledgements	xxiii
	Maps	xx
	Abbreviations	xxiv
1	Early life and education, 1806–32 *Janet H. Robinson*	1
2	Courtship, curacies and marriage, 1826–56 *Janet H. Robinson*	23
3	The Hardwicke years, 1856–85 *Mark Robinson and Janet H. Robinson*	35
4	Man of the cloth *Mark Robinson*	59
5	The antiquarian and historian *Mark Robinson*	74
6	Webb's observations of earthquakes *Roger M.W. Musson*	85
7	Clerical astronomers *Allan Chapman*	101
8	Webb's telescopes *Robert A. Marriott*	121
9	Webb's observing notebooks *Peter D. Hingley*	145
10	Webb and the Moon *William Sheehan*	152

11 Webb among the planets 165
Richard M. Baum
12 The comets observed by Webb 181
Jonathan D. Shanklin
13 Webb's observations of the Sun 194
J.C.D. (Lou) Marsh
14 Double stars, garnets and rubies 200
Robert W. Argyle
15 Celestial objects for common readers 215
Bernard Lightman

Appendix 1 The Webb Society: its history and activities 235
Robert W. Argyle
Appendix 2 Bibliography of Webb's published works 241
Jaspreet Gill and Bernard Lightman

Index 252

Illustrations

Revd Thomas William and Mrs Henrietta Webb Frontispiece
Maps showing locations in Herefordshire xx
Maps showing locations in Herefordshire xxi
Sketches of insects 1
The rectory of the parish of Tretire with Michaelchurch 3
Webb's diary for 1827 6
One section of Webb's diary 6
Magdalen Hall, by J. Fisher, 1827 13
The Sheldonian Theatre in the mid-nineteenth century 15
The Radcliffe Observatory, Oxford, in 1814 17
Electrical apparatus by Michael Faraday 19
A page of Webb's diary, showing his shorthand 24
St Weonard's Church, Herefordshire 27
T.W. Webb 37
Arthur Cowper Ranyard in 1879 37
Sketch of donkey's head, by T.W. Webb 40
Hardwicke Vicarage in 1867 43
Drawing room door, painted by Henrietta Webb, at Hardwicke Vicarage 48
A sketch by T.W. Webb 51
The cover of the sale catalogue of Hardwick vicarage, 1885 55
Thomas William Webb – *a carte de visite* 62
Holy Trinity Church, Hardwicke, in 1867 66
Holy Trinity Church, Hardwicke, probably painted by Mrs Webb 66
A view of Hereford Cathedral from the north-east, 1831 69
Webb's sketch of a tump and moat 75
Revd John Webb 78
Brampton Bryan castle 81
Map showing the locations mentioned in text 95
Intensity map of the Abergavenny earthquake of 16 January 1883 99
Revd Edward Lyon Berthon 112
Revd T.H.E.C. Espin 113
Chromatic and spherical aberration 123

A Tulley refractor of the mid-nineteenth century 127
The 3½-inch refractor by Dollond 127
A 9⅓-inch With–Berthon reflector 128
A 'Romsey' observatory 128
George Henry With 135
Sir William Huggins 135
The inscription inside the front cover of Webb's notebook 146
Donati's comet, 1858, in Webb's notebook 150
The crater Schröter, in Webb's notebook 153
Webb's notebook, Volume 2 155
The twin craters Messier, in Webb's notebook 158
Helicon A and Helicon, in Webb's notebook 160
Observations upon Venus, Webb's notebook 168
Venus with 'light brushes', observed by J.H. Mädler 172
Thirty drawings of Mars, in Webb's notebook 174
Drawings of Saturn, in Webb's notebook 178
Donati's comet, 1858, in Webb's notebook 186
Comet Hyakutake, 1996 188
Drawings of the head of Donati's comet, 1858, in Webb's notebook 189
Drawings of the head of comet Hale–Bopp, 1997 190
Drawings of the Great Comet of 1861, in Webb's notebook 191
The Great Comet of 1882, in Webb's notebook 193
Diagram of a sunspot, in Webb's notebook 196
Observations upon the Sun, Webb's notebook 197
The first page of Webb's last observing notebook 202
The last observing entries made by Webb 203
A typical page in Alexander Jamieson's star atlas 206
The light curve of the variable stars S Orionis 211
The wide multiple star Web 10 212
The Great Nebula in Orion, as drawn by Bond in 1847 228
Trouvelot's depiction of the Great Nebula in Orion in 1874 229
The Great Nebula in Orion, drawn by Muschamp Perry in 1881 229
Kenneth Glyn Jones, Founder of the Webb Society 236

Picture credits

Royal Astronomical Society (frontispiece and pp. 37, 40, 51, 112, 113, 146, 150, 153, 155, 158, 160, 168, 174, 178, 186, 189, 191, 193, 196, 197, 202 and 203); Hereford City Library (pp. 6, 24, 55, 69 and 135); Hereford Cathedral Library (p. 78); Oxford Centre for Local Studies (pp. 13, 15 and 17); Private collection/photograph National Portrait Gallery (p. 62); Peter Denton (pp. xx and xxi); T.C. Platt (p. 188); D. Graham (p. 190); Dr A. Catterall (p. 212); Brenda Glyn Jones (p. 236).

Contributors

Robert W. Argyle	Officer at the Institute of Astronomy, Cambridge; joined the Webb Society in 1968, and in 1970 became Director of the Double Star Section – a post which he still holds; appointed President of the Society in 1990; joined Commission 26 (Double Stars) of the International Astronomical Union in 1994; appointed editor of *The Observatory* in 1996; editor of *Observing and Measuring Visual Double Stars* (2003).
Richard M. Baum	Member of the British Astronomical Association, and Fellow of the Royal Astronomical Society; published papers in many astronomical journals. Co-author (with William Sheehan) of *Le Verrier's Wild Geese: the Romantic Quest for the Intra-Mercurial Planet Vulcan* (1996) and *In Search of Planet Vulcan* (1997). Contributed to *Images of the Universe* (ed. C. Stott, 1992) and *The Observational Amateur Astronomer* (ed. P. Moore, 1995), and to biographies of forty astronomers in B. Lightman's *Dictionary of Nineteenth Century British Scientists* (2004).
Allan Chapman	Fellow of Wadham College and lecturer in the Faculty of Modern History, Oxford University; lectures extensively and widely on the history of astronomy and science; author of several books, including *Dividing the Circle: The Development of Critical Angular Measurement in Astronomy, 1500–1800* (1995), *The Victorian*

	Amateur Astronomer: Independent Astronomical Research in Britain, 1820–1920 (2001), *Gods in the Sky* (2002), and presenter of the Channel 4 television programme of the same name.
Peter D. Hingley	Librarian of the Royal Astronomical Society; author of several articles in local history and scientific journals; co-author of *A Far Off Vision: the Autobiography of Edwin Dunkin* (1998).
Bernard Lightman	Professor of Humanities, York University, Toronto, Canada; current editor-in-chief of *Isis*; has researched and written extensively on the social history of science and its popularisation; edited *Victorian Science in Context* (1997); produced *Dictionary of Nineteenth Century British Scientists* (2004).
Robert A. Marriott	Amateur astronomer, writer and lecturer; Director of the Instruments and Imaging Section and Curator of Instruments, British Astronomical Association; formerly a dealer in rare books, and now a freelance editor of books on science and history; author of numerous papers in astronomical journals.
J.C.D. Marsh	Former Head of Astronomy and Director of the Observatory at the University of Hertfordshire; after retiring published papers on lunar photography and radio emission from the solar corona; always known as Lou, but never revealed the origin of this name; died in 2004.
Patrick Moore	Presenter of the television programme *The Sky at Night* since 1957; author of numerous books; one of the best loved astronomical popularisers in the world.
Roger M.W. Musson	Head of Seismic Hazard at the British Geological Survey; compiler of the standard catalogue of historical earthquakes in Britain; interested in the history of seismologists and seismology.
Janet H. Robinson	Author of several articles in local history magazines; currently working on the transcription and increased availability of the diaries and letters of the Webb family.

Mark Robinson	Retired from the computer industry; now researches and publishes local history; author of the entry on T.W. Webb in the *Oxford Dictionary of National Biography* (2004).
Jonathan D. Shanklin	Head of the Meteorology and Ozone Monitoring Unit at the British Antarctic Survey; Director of the Comet Section of the British Astronomical Association and of the Society for Popular Astronomy; wide-ranging interest in natural history.
William Sheehan	Amateur astronomer and historian of science; author of several books, including *The Planet Mars* (1996), co-author (with R. Baum) of *Le Verrier's Wild Geese: the Romantic Quest for the Intra-Mercurial Planet Vulcan* (1996) and *In Search of Planet Vulcan* (1997), co-author (with T.A. Dobbins) of *The Epic Moon: A History of Lunar Exploration in the Age of the Telescope* (2001), and co-author (with J. Westfall) of *The Transits of Venus* (2004).

Foreword

Thomas William Webb has an honoured place in the history of astronomy. He made no great theoretical advances, and he made no spectacular discoveries, but for a long period in the nineteenth century he exerted tremendous influence. Obviously I never met him, because he died in 1885; but I have met people who knew him, and it is clear that he was a very exceptional person. He had many friends, and it is unlikely that he had any enemies.

Webb is best remembered today for his classic book, *Celestial Objects for Common Telescopes*. The first edition was published in 1859, and the sixth (edited by T.H.E.C. Espin) in 1917. It was unlike anything else – the only possible comparison being W.H. Smyth's *A Cycle of Celestial Objects* (1844) – and it became the standard work for amateurs; and professional astronomers also found it useful. But though this was undoubtedly Webb's most important work it was by no means the only one, and he was amazingly versatile. This is why the present book is so valuable. It deals with all aspects of his career – not forgetting the fact that he was a dedicated and hard-working clergyman. What emerges is a picture of a truly great Victorian.

Webb was not the only science populariser of his time – others of note were Admiral W.H. Smyth, Richard A. Procter and Robert Stawell Ball – yet on the whole he had more impact than any of them, because he was essentially practical; and *Celestial Objects*, in particular, was a book to be taken into the observatory rather than left indoors lying on a shelf. Although the last edition appeared so long ago, it is still often used today.

This is the first book to pull together all of Webb's activities. It will appeal to scientists and non-scientists alike, and is a fitting tribute to a man who, during his lifetime, contributed so much.

Sir Patrick Moore

Editors' preface

I have had it in hand for a long while, having had so many things to attend to. T.W. Webb to A.C. Ranyard

This book has been a long time coming. In 1890, five years after Webb's death, T.H. Foulkes, Chaplain to Her Majesty's Forces, in a letter to the British Astronomical Association wondered 'why [Webb's] life and letters have never been published'. He suggested that if only his letters were collected and published, 'every Astronomer in the world would get a copy, out of the love we all bear to the author of *Celestial Objects*'. S. Maitland Baird Gemmill replied:

> Something in the same idea occurred to myself, though what I thought of did not include a memoir, but only a collection of papers and observations. I wrote to Mr Espin on the subject, who replied that as for the unpublished observations of Mr Webb, they were in his (Mr Espin's) possession, and he was getting them into print as rapidly as opportunity offered, and as for the published papers, he feared the idea was impracticable. After this I did not persevere with the proposal, but have kept the idea in my mind ever since. The work would be a worthy memorial of one who did so much to advance the cause of astronomy. I hope the suggestion will be discussed.

Three years later, Gemmill again wrote in the BAA *Journal*. He said that after his letter appeared he was in communication with Mr Ranyard, Mr Espin and others, all of whom gave approval and encouragement. He was loaned some letters and made notes and collections, but ill-health and a 'long series of untoward circumstances' interrupted his endeavours. He was therefore making another start, begged members to render assistance and avowed that any communications would be attended to at once. Sadly, that was the last mention of Gemmill's projected work.

In 1901 Arthur Mee produced a loving Memoir of Webb in the *Cambrian Natural Observer*, but it was to be another thirty-five years before the subject was raised again. In 1936 Alfred Noel Neate pointed out to BAA members that since it was fifty years since Webb died, it was hardly likely that a complete 'Life and Work' would ever be published; but he suggested that those members who had notes, letters, papers, observations, reminiscences or anything relating to Webb should come forward so that the Association 'may have as permanent a record of Mr Webb's life and work as it is now possible to secure'. Neate himself offered a few notes relative to visits he had made to Hardwicke and Mitchel Troy in 1912 (see Chapter 3). Nothing seems to have resulted from this plea, and silence again fell until 1950.

At that time there was correspondence between a Mr Moore FRAS and the librarian of Hereford Cathedral. Mr Moore had put together a short article for publication in an astronomical magazine, but intended to produce a longer piece to be published in 1985, the centenary of Thomas Webb's death, but ill health had overtaken him, and his visit to the archives in Hereford apparently never came about. (This was probably Eddie Moore, of the Webb Society, who died in 1985.)

That might have been the end of any attempt to commemorate Webb had we not bought his old vicarage at Hardwicke in 1986. We are not astronomers, so we knew nothing about him; but, being strongly interested in local history, we discovered that the Revd Francis Kilvert mentioned the 'Webbs of Hardwick' in his diary. Consequently we began to investigate the history of one who we thought would turn out to be a worthy, but probably dull, Victorian clergyman. After visiting the library of the Royal Astronomical Society and sampling some of Webb's letters and notebooks, we realised that we were wrong. Peter Hingley, the librarian, encouraged us to hold a Webb Weekend at Hardwicke in November 1990, during British Astronomy Week. The willingness of Patrick Moore and other astronomers – including three of the authors of this book (R.W. Argyle, P.D. Hingley and R.A. Marriott) – to attend and to lecture surprised us, but in talking with them we began to appreciate the affection in which Webb was still held. We were astonished to realise that a scientific work – Webb's *Celestial Objects for Common Telescopes*, first published in 1859, and reprinted as late as 1962 – was still being used, providing inspiration for modern amateur astronomers.

The pursuit of the charming, gentle and meticulous Thomas William Webb became more than a casual pastime. Indeed, during quiet winter days when we sat in the vicarage transcribing diaries or making notes, the atmosphere became so redolent of Thomas and his wife Henrietta that if they had walked down the hall and into the library there would have been no sense of surprise.

We were fortunate that Hereford City Library, Hereford Cathedral Library and the County Record Office, as well as the Royal Astronomical Society, held archives which were a modest cornucopia of primary sources: letters, wills, notebooks and early personal diaries. Our progress was extremely slow – partly because we were fully involved in running a guest house, and also because the diaries and many of the letters required careful transcription. Several years later, Peter Hingley suggested (notwithstanding the apparent fate that seemed to attend eager Webb biographers) that a book celebrating Webb's life and work should be attempted; and if we would write a biography, he would invite others who could illustrate various aspects of Webb's contributions to astronomy. His assemblage has been both prestigious and masterly.

It has fallen to us to become not only contributory authors but, through a 'long series of untoward circumstances', novice editors. The astronomers and academics – amateur and professional – who have contributed to this book, have been both generous and patient, answering our queries and giving their advice. We have edited their expertise with a delicate touch. And though the production of a multi-author work has occasionally been infuriating it has given us much pleasure to work with them, and we are extremely grateful to each and every one.

Every contributor has used Webb's own words wherever possible, for he wrote with liveliness and wit. Indeed, most of the epigraphs are his own words (and come from *Celestial Objects* unless otherwise referenced). We all realise that what we have produced is not exhaustive. Webb was such a prolific writer of letters and articles that we could not hope to make the work complete. For instance, no overseas repositories have been sought, let alone tapped. The sources which we have used to describe his life in Herefordshire contain much more for local historians than we could make available here. However, we hope that astronomers will enjoy the biographical details and find that their picture of the author of *Celestial Objects* has been enriched, and that the general reader will discover, as we have ourselves, much in the astronomical chapters to inform and delight them.

Janet and Mark Robinson
Hay-on-Wye
September 2005

For those who would like a reminder of the area in which Webb lived and worked, we proffer the maps on the following pages.

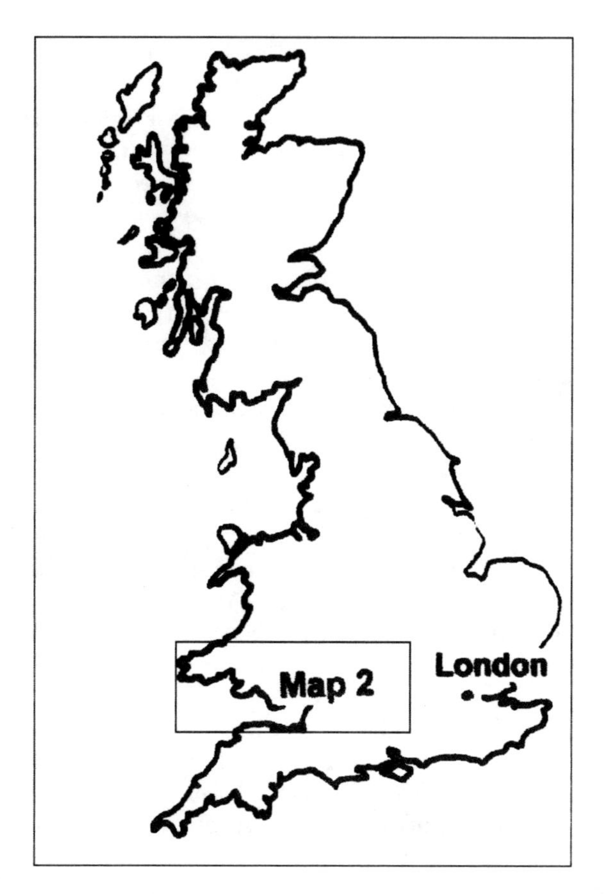
London
Map 2

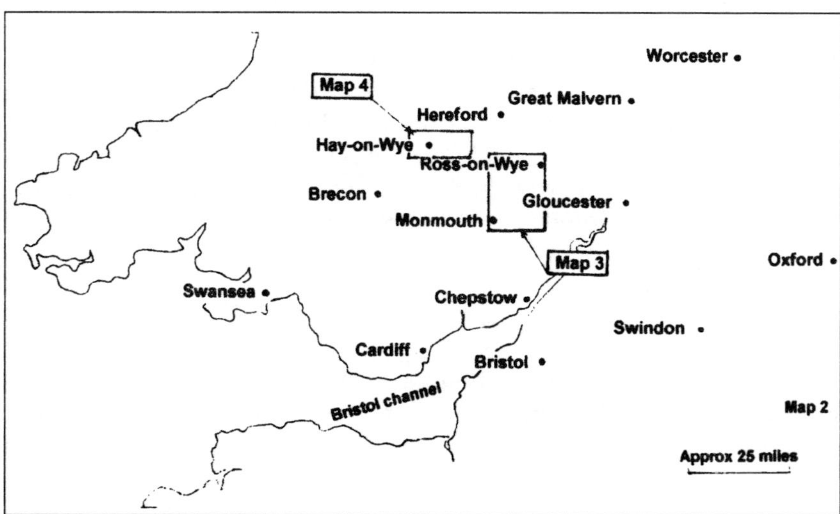
Worcester
Map 4
Great Malvern
Hereford
Hay-on-Wye
Ross-on-Wye
Brecon
Gloucester
Monmouth
Map 3
Oxford
Swansea
Chepstow
Swindon
Cardiff
Bristol
Bristol channel
Map 2
Approx 25 miles

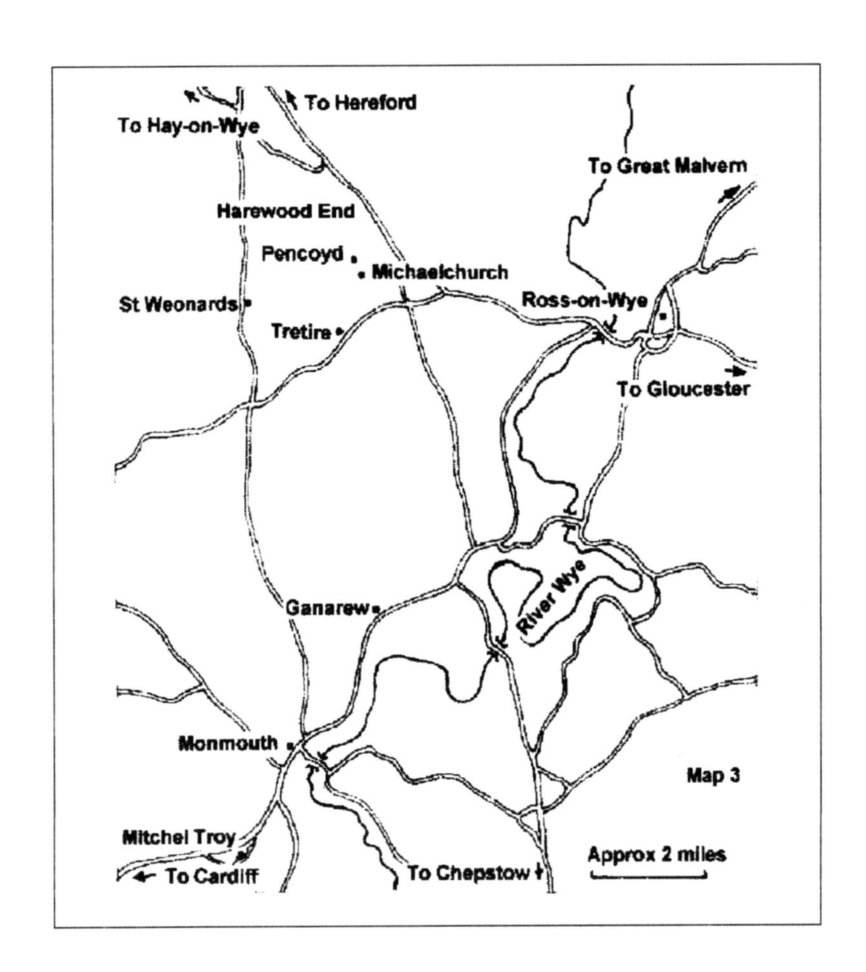
To Hay-on-Wye
To Hereford
To Great Malvern
Harewood End
Pencoyd
Michaelchurch
St Weonards
Ross-on-Wye
Tretire
To Gloucester
River Wye
Ganarew
Monmouth
Map 3
Mitchel Troy
To Cardiff
To Chepstow
Approx 2 miles

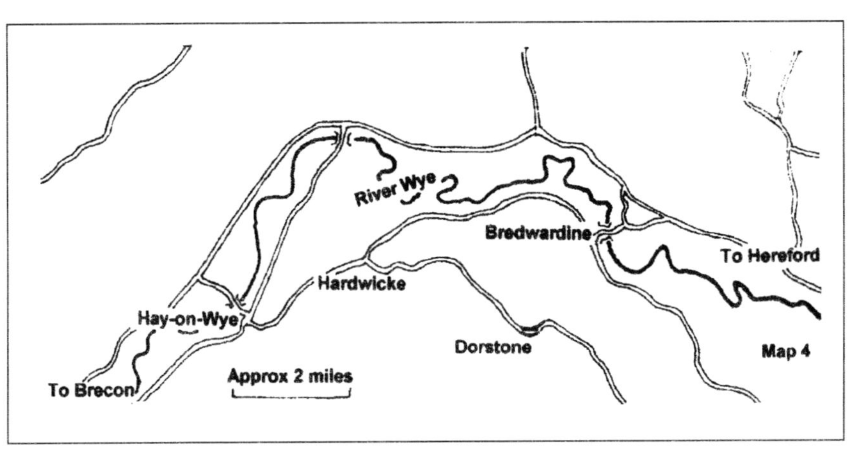
River Wye
Bredwardine
To Hereford
Hardwicke
Hay-on-Wye
Dorstone
Map 4
To Brecon
Approx 2 miles

Acknowledgements

We are most grateful for the expert help and generosity of several librarians and archivists, particularly Robin Hill at Hereford City Library, Peter Hingley and Mary Chibnall of the Royal Astronomical Society, Joan Williams, until lately at Hereford Cathedral library, and Sue Hubbard, recently retired from Hereford Record Office. Staff at the British Library and the Bodleian Library have also willingly steered us to our objectives.

The Royal Astronomical Society has been most generous in allowing us to reproduce so many images from the archives, and we acknowledge similar generosity from Hereford City Library and the Oxford Collection of Local Studies.

Members of staff at Gracewing have willingly endured our many queries, and we have leaned heavily on the astronomical and editing expertise of Robert Marriott. We are also indebted to him for producing the index. Peter Denton kindly created the maps at the beginning of the book. To all of these, our grateful thanks.

We also wish to acknowledge the support and encouragement of many friends who have listened patiently to the fruits of our researches.

Finally we would like to thank each other: one for the dogged determination which ensured that the manuscript reached a publisher; the other for the patient logic which has taken it over the last hurdle.

Janet and Mark Robinson

Abbreviations

Three of the manuscript sources used frequently in the notes and references are abbreviated throughout the book as below, followed by the date:

Diary	T.W. Webb, *Diary 1826–37* and *1840*, MS in Hereford City Library.
Webb to Ranyard	A collection of 331 letters written between 1858 and 1885 from T.W. Webb and Henrietta M. Webb to Arthur Cowper Ranyard and preserved in the archives of the Royal Astronomical Society.
Kilvert, *Diary*	W. Plomer (ed.), *Kilvert's Diary: Selections from the Diary of the Revd Francis J. Kilvert, 1870–79* (Jonathan Cape, London, 1938–40).

Chapter 1

Early life and education, 1806–32

Janet H. Robinson

My father was curate to the Revd. Thos. Underwood, Rector of Ross at the beginning of the present century (during which time I was born, at a house called The Cottage, on the Walford road, Dec. 14 1806).[1]

Early life and ancestry

August 7th, 1817: I placed a spider on a bent stick, inserted in a flat piece of wood, which was kept at the bottom of a basin full of water by means of stones. The spider, after many fruitless attempts to escape, threw out a thread of about two feet in length which attached itself to a chair. The spider felt with its forelegs to ascertain whether it was firm, and it proving so, passed over it. See Kirby and Spence's Introduction to Entomology Vol.1 p.414.[2]

Thus, in neat copperplate handwriting, Thomas William Webb made his first entry in a notebook entitled 'Observations upon Natural Phenomena. Part I'.[3] He was ten years old. In a second notebook headed 'Entomological Remarks' he made a fair edited copy of his notes, illustrating them with small detailed sketches, often in watercolours. Loose inside this second book is a small hand-made paper measure for estimating the size of insects.[4] He made entries every two or three days, and sometimes daily. He also recorded the first snowdrops, and noted the arrival of the swallow and the cuckoo, as well as odd occurrences like 'Papa coming over May Hill saw a rabbit chasing a stoat'. On 5 January 1818 the first astronomical observation appeared:

> Saw a meteor about the breadth of the semi-diameter of the moon, about 5h.30m P.M.: it passed from NE to SW across the zenith, its observed time was about 3s. It may be estimated at 60° N. Decl. when in Long. 15°.[5]

Two weeks later he wrote about a mountain on the Moon, and apparently drew a diagram in another notebook (now lost):

> Jan. 17th Observed mountain in the moon (9) at 'a'. A streak issued from it at 'b'. Another mountain was seen shining like a star at 'c'; a third at 'd'. From d to e the edge of the shadow was very irregular; the points f & g looked like craters – the whole surface was strongly freckled.

Thomas made these notes at Tretire – a hamlet near Ross-on-Wye, in Herefordshire. The careful observation and the recording of the experiment and his sources were certainly the result of his father's teaching, for John Webb had, as his son later wrote of him, 'a mind of no common accuracy and candour.'[6] It was undoubtedly he who inculcated and encouraged the qualities which made Thomas such an accurate and meticulous observer.

John Webb originated, on his father's side, from a family of Wiltshire and Dorset gentry. An ancestor, who was a Catholic, fought with such distinction on the Royalist side in the Civil War that he was made a baronet. Indeed, Thomas Webb himself became official head of that family when the seventh baronet died childless in 1874.[7] However, the cadet line from which John sprang directly was almost certainly descended from Cromwell's principal Surveyor-General of the Crown Lands.[8] John's father, William Webb (1736–1791) was a druggist in Castle Street, London, but he was also a talented academic, being 'a mathematician, political writer, linguist and Hebraist'.[9] John's mother was Anne Sise, 'the daughter and heiress of James Sise the medical officer to the Aldgate dispensary',[10] who was descended from Huguenot refugees. A bachelor uncle, Thomas Webb of Hoxton, who was Surgeon to the First Regiment of Foot and later had an extensive practice in London, left property and money to John when he died in 1812.[11] William and Anne's sons followed varied careers, one being a lawyer and another an engraver to George III, while John was the only cleric.

John Webb was born in 1776, and went to St Paul's School, where he became Captain of the school and proceeded, as a Pauline Exhibitioner, to Wadham College, Oxford, where he took his BA degree in 1798 (and his MA four years later). The following year, and before his ordination, he married Sarah Harding at St Clement Dane, Westminster. It may well have been a love match, for John was only twenty-two and had yet to make his way in the world. Sarah had a marriage portion of £6,500, the interest of which, even at 2½ per cent, would have been twice a curate's stipend. The Hardings were a landed family, having property in Warwickshire and London, though Sarah's father, like John's, was of the cadet branch. In 1800 John became a curate in the Lichfield diocese. In the following year he went to Ripple, Worcester, where Sarah gave birth to a girl, Anne Frances. Their son Thomas William Webb was born on 14 December 1806, when John was curate to the Revd Thomas Underwood at Ross-on-Wye. Two children was not a vast number for those prolific days. Perhaps John Webb spent too many night hours studying or in conversation with his friends, but they had a son to carry on the family line. Sadly, their daughter died the following spring, in her eighth year, and Thomas William became an only child.

The Webbs left Ross-on-Wye in 1809, and John held various appointments in the Midlands. In 1812, when he was thirty-six, he became rector of Tretire with Michaelchurch, five miles from Ross. The handsome stone rectory – reached up a narrow lane leading from

The rectory of the parish of Tretire with Michaelchurch.

the road near St Owen's Cross – had been built in the early years of the eighteenth century. Around it were the undulating fields of the scattered farms that made up the parish. Further to the west stood the high, wild hills of the Welsh border. It was not a prestigious appointment. The parish was entirely rural, containing about a hundred and twenty souls, and the stipend was £180 per annum. Nonetheless, John held this living until his retirement in 1859, when he installed a full-time curate who ran the parish until John's death ten years later at the age of ninety-two. John's clerical talents were recognised locally, for he was made a minor canon of Gloucester Cathedral in 1820 and was given, by that Dean and Chapter, a living in Wales – that of St John and St Mary, the parish church of Cardiff, which he held *in absentia* from 1821 to 1863.

His other activities revealed him to be a considerable scholar in various fields. His proficiency in Norman French was 'remarkable' and he was sometimes asked to give evidence in the law courts on the interpretation of early documents. In 1819 he was elected a Fellow of the Society of Antiquaries, and thereafter contributed papers to that society on esoteric and varied subjects. These researches took him to the Bodleian Library in Oxford and to London, so that he was hardly buried in his rural parish. Some of his papers were privately printed,[12] but he also edited, for the Camden Society, *A Roll of the Household Expenses of Richard de Swinfield ... 1290*, which was published in two volumes in 1854/55 and is still used as an historical source. Research into the history of the Civil War in seventeenth-century Herefordshire engaged him throughout his years in the county.[13] He adapted part of Haydn's *The Seasons* for performance at the Birmingham Music Festival in 1834, and wrote the libretto for the oratorio *David*. This work, composed by his friend Chevalier Newkomm, was performed at the same festival, and was also received with acclaim in America.

With his clerical appointments, some inherited money and the dowry of his wife, John had a reasonable income with which to provide for his small family. To augment it, and to provide company for Thomas, he took a few pupils – the sons of neighbouring clergy and local gentry. There is no evidence that Thomas Webb was ever sent to school. Maybe John considered the local schools inadequate; and perhaps the thought of sending Thomas away to a harsh boarding school régime may have been anathema to both his parents doting on their one remaining child.[14] Whatever the reason, Thomas William received all his education from his father until he went to Oxford. It is clear that he had the usual classical education plus mathematics, French, German and some Hebrew. Arthur Ranyard, in his obituary of Webb,[15] states that Thomas was a painstaking, carefully precise boy, preternaturally old and studious – an impression which one feels can only have come from Thomas's estimate of himself.

The only other evidence of Thomas's early education is the surviving notebooks quoted at the beginning of this chapter. They show a growing and informed interest in the natural world, and of astronomy in particular. The pace of the young Thomas's life gave time and opportunity for the close and detailed methods of study that were taught him by his father. Besides continuing to write up his notebooks he kept separate accounts of his astronomical observations from 1826.[16] There is also extant a most detailed record of the weather, with daily entries for the years 1825–26.[17]

Thomas also kept a diary. It is possible that this was a lifelong habit, but the only diaries currently known are those which he kept from 1826 to 1840.[18] The first year was written in a small worn pocket book, while the rest were recorded on sheets of foolscap paper carefully folded to make sixteen columns and then usually cut, probably when completed. Each foolscap page recorded about three months in the earlier years, though from 1834 the entries are less detailed, no doubt as his clerical duties occupied more of his time. The handwriting is tiny – so tiny that reading it usually requires a magnifying glass. Some words and a few whole phrases are now indecipherable. The outer folds are worn and faded as if the owner kept them carefully in his pocket. It is almost entirely from this source, besides a few random letters, that the life of Thomas from his nineteenth to his thirty-fourth year is known.[19]

Thomas's diary is laconic and factual. He appreciated the wild countryside around him, but his remarks are confined to noting the view and describing it as 'glorious'. The weather, however, was rigorously recorded, and he used words with accuracy and originality. He spoke of a 'rattling rollocking wind', a 'most wonderful, strange crisping fog' and 'most beautiful cirrostratus [which] after sunset broke into a thousand bars and feathers, "An Angel's wing"'. He observed the changing patterns: the rain-clouds rushing in on the predominant south-westerly wind, and the harsher weather that comes from the north-east. Examine even entries within one month – January 1826, when the extant diaries begin – and his notes and asides on the weather and its effects inform the whole and provide us with glimpses of the conditions in which he lived. On 1 January there was a 'Sudden and extraordinary thaw – heavy rain'. Next day there was a 'Very high flood up to the ditch in Tumps – Pa could not go [to Birmingham] for fear of floods'. Thomas himself walked through deep mud to see a parishioner at Aberhall. Three days later there was a 'Sudden and thick fall of snow – snowed all day and lay thick'.

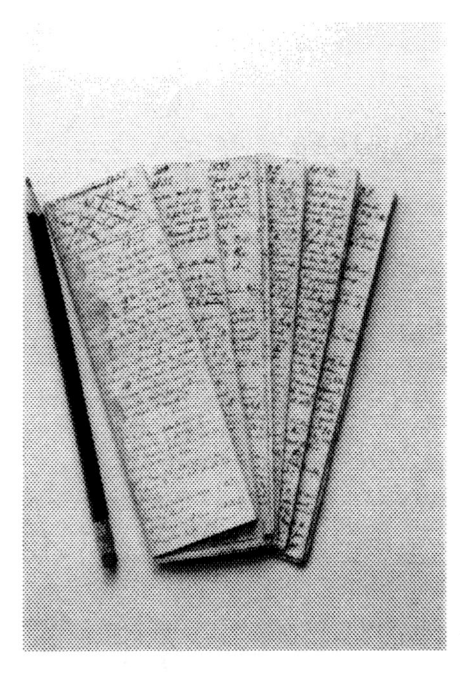

Webb's diary for 1827.

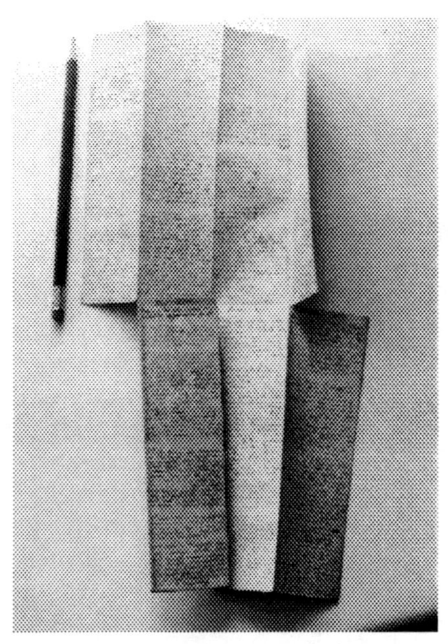

One section of Webb's diary, showing the complete page unfolded.

Hurdles had to be placed along the path to the privy to keep it clear. It was 'intensely cold. [The river] Wye had been frozen on the 12th and was expected to be covered on the 13th. House full of robins. Therm[ometer]. at 10h pm down at 12deg.!!!'. By the end of the month, walking was 'filthy greasy work – whole fields under water'.

The diary contains almost no social or political comment, although Thomas recorded anecdotes, ghost stories and dreams. The writing style is fundamentally different from that of his observing notebooks, in which the entries are detailed, extended and accurate and there is no possibility of misconstruction. When recording the daily round he made abrupt changes of subject and, since he also used the common Victorian convention of a short dash placed on the line – which is treated as a comma, dash, full stop or ditto mark – the entries occasionally have a ring of Alfred Jingle.[20] For example:

> 1826. 14 February: Planting laurels but very wet. Penygate – Hollis Bridge – Mamma's wrist. Sent to Mr Gyttings who came from Pencraig. Mamma in a great alarm for fear of an operation – Taking up 3 Ash Trees. Mrs Mary Hill very poorly and low. Neck of Mutton sent for by Blewit – Sarah Gundy came over – Schroter.

The family spent over half the year – mainly from July to February – in residence at Tretire, where Thomas assisted with the agricultural tasks on the glebe land: haymaking, harvesting the autumn fruit, and planting and tending the garden. There were the usual problems of country life: workmen failed to turn up, wasps' nests had to be removed with squibs or boiling water, and there were pigs in the orchard or sheep in their fields 'devouring turnips, 'taters, beans, parsnips etc'. Most afternoons he rode his dearly loved and spirited mare, Pretty Polly, for about six to ten miles through lanes and hamlets with evocative names like Llwyn Ddu, Treberran and Trewathen Pool, where there were 'great waves and a man shooting moorhens'. He took pride in Polly's performance, and frequently recorded her time over distance, adding 'Bravo Pretty Polly!!!'. Occasionally he fell, though with little damage. He never told his mother, fearing even to rub liniment on his bruises in case she smelled it. Hers may have been natural concern, but one suspects that she fussed.

Thomas read much – mainly astronomical works, but also Kant, *Blackwood's* magazine and Sir Walter Scott, amongst others. He sketched, making his own crayons, and he made and sailed a model yacht on the pool near the rectory. Such activities were lifted from mere boyish activity by his curiosity and observation. If crayons were mixed with water, he wrote, they were too soft, and with milk, too hard. Sailing the boat became a scientific experiment as he altered the sails and recorded its performance.

Social life at Tretire was limited. Thomas rode into Ross on errands about once a week, and talked with acquaintances and tradesmen. When he was invited to dine at the houses of the local gentry he might note the conversation, but not the food.[21] They had visitors only rarely, the most frequent being the various clergy who were deputed to take the services when John Webb was away.

The household maintained three indoor servants and the outside man, Thomas Hill, and there were other part-time cleaners and agricultural contractors. Hill was with the family for many years, and since Thomas Webb frequently reported his conversations and stories in his diary he must have been as much companion as servant. The maids and cook, however, changed frequently. Either they did not like the isolation of the rectory, or Mrs Webb was an erratic and capricious employer.

Sarah Webb seems to have led a secluded and somewhat solitary life. Her husband was often absent, and she did not have a faithful servant who could have been a female confidant. Thomas did much to entertain her, accompanying her on walks, driving her out in the gig, or walking beside her when she rode on the family donkey. They must have spent many evenings alone together, and there seems to have been much laughter, although banter that made them 'laugh to splitting' is now incomprehensible – lost between Thomas's brevity and our ignorance of jokes peculiar to the family. He appreciated her sense of humour, often quoting her remarks in his diary:

> A letter from my Aunt. Ma said of Mrs R. – boasting of being a blue [stocking] – a blew as Mrs R. called it – 'Pray, Ma'am, are you a stone-blew, or a powder-blew?'[22]

He happily reported wit directed against himself:

> I saying 'I did not like to go anywhere to make a fool of myself,' M. said 'I need not take the trouble, other people w'd make a fool of me'. Bravo!'[23]

Is the 'Bravo!' added with a little condescension? Occasionally when Thomas wrote about his mother his tone was somewhat patronising, but then what young man does not sometimes think his mother is a fool?

When he spoke of his father there was no hint of censure. When John Webb was at Tretire they walked and rode together, played backgammon, chatted about local history or learned treatises and discussed the news and anecdotes that John brought back from the wider world. One suspects that while Sarah Webb enjoyed the gossip from London and Gloucester, she did not share the academic interests of her husband and son. Spare a thought for the lady when the former spent the whole evening poring over historical documents and the latter was outside viewing the Heavens!

Gloucester

> *Took leave of my dear Pretty Polly, had my 3ft reflector put in chaise & off to Glo'ster.*[24]

From 1826 to the beginning of 1831, and possibly for several years earlier, the Webbs spent much time in Gloucester, twenty miles east of Tretire. Besides the telescope, a maid and a basket of cats were all packed into a chaise, and the family would rattle off to a rented house so that John could more easily perform his duties at the cathedral. Unlike some, he was a conscientious canon, and his residence for several months of the year would have entitled him to a share in the capitular income of the cathedral.[25] Gloucester was at that time a substantial city of about 10,000 people, and Thomas records 'bustling about' and uses the word 'hubblebubble', which suggests he was conscious of the crowded streets and urban activity.

In Gloucester he lived the life of a young gentleman, though this was hardly a social whirl. He rode most afternoons in the rural environs of the city, occasionally accompanied by his father. Churchgoing, too, remained an unaltered part of his life. He usually attended Matins at the Cathedral in the morning – a service that could last two and a half hours with a long sermon – and he sometimes went again in the evening. Services were social occasions; the congregation as important as the worship: 'Spa church evening ... Mrs. Gladstone's ear trumpet fell into the passage. "Thus saith the Lord" Handel ... Mr. Buckle had a bad headache. Saw Mrs. Sikes evening very much reduced.'[26] Thomas associated with the staff of the Cathedral and enjoyed the gossip and interesting conversation. Some of his father's friends, such as the Revd J. Bishop, were extremely interested in astronomy and science. Thomas used his library, and sometimes read there half the morning.

Social life with his mother was more mundane. He accompanied her on morning visits where the company was almost exclusively female and the life of Gloucester society was raked over in age-old fashion. They called on 'Miss Phillpotts & had the bother of seeing & hearing the old lady at full length, her daughter being out ... Miss P. called on us at tea & talked us all to death.'[27]

In the evenings he sometimes played piquet with his parents, but there were more frivolous occasions. He was invited to Mrs Buckle's rout with supper, but makes no comment on it. He was, not unnaturally, shy in female company, and perhaps he did not dance. He accompanied his father to dinner with the latter's friends, and Mrs Webb rarely went with them. When the Three Choirs Festival was held at Gloucester, the Webbs made a special visit, staying about a week.[28] The *Messiah* was often performed, and Thomas wrote of song recitals including 'Papa's song, set by Callcott & sung by Vaughan'.

Thomas Webb was, however, first and foremost an astronomer. Observation and telescope-making were the ground bass and continuum of his life. Most of his diary entries at both Tretire and Gloucester make mention of them. When riding he 'thought of convex speculum'. Out walking he was thinking 'over and over again of Cassygrain, Dick, Newton and so on'. In Gloucester there was time and facilities for telescope-making, and people with whom he could discuss his passion. On 19 May 1826 he says that he 'Commenced regular account of observations'. This was his first notebook specifically on the Moon.[29] Mainly he observed alone, but occasionally he had company. On 11 March 1826 he was

> With Pa looking at Sun's spots, on paper and through smoked glass. Moon beautiful in the evening. Mr Bishop called and we observed together, planets, stars &c. he praised my glass much: promised me his dark glass.

Sarah Webb apparently had very good sight, and Thomas, in his account of observing the satellite of Venus in 1826, said that she could see the satellite with the naked eye. (This 'satellite', however, does not exist.)

He frequently spent the greater part of a day with local tradesmen who could make parts for his telescope and stand. One of these, Charles Bonnor, was a brazier, brass founder and tinplate worker who worked in Westgate Street.[30] White and Moss were whitesmiths, and they too were often persuaded to stop what they were doing and make a nut or screw for a particular purpose. Thomas was always curious and observant, and while visiting the tradesmen he often commented on other work and processes in the workshop. Although he badgered them incessantly – often visiting the workshop three or four times a day and late at night – they seemed to have had a liking for the unassuming and friendly young man:

> 1827 April 5. Went to White: c'd get but little done after waiting all morning it was so hard his tools w'd hardly touch it – at last put the stone against the face of it and kept it there by puppet screws while the metal revolved – then came back & rode to Upton ... In town again w. Pa. went to White – found he had done one by making it revolve & screwing the stone against it – a pretty good one. We did the other using water, that the speculum might not get too hot; when dry ground, it was soon too hot to touch, & actually assumed a copper colour. Some sand holes in it. When done, went to Bonnor – out.[31]

Thomas thought nothing of 'grinding away' at a speculum for two or three hours at a time, and he had to design and have made the tools that would do the job, experimenting with various materials like emery and colcothar.[32] He usually worked at home, but also used the facilities of the workshops:

> Went to Moss: found a nice apparatus of lapidary's wheel: wh. had no occasion however to be spun round, only turned 45° at a time, so I was unable to sit at the work. made a polisher & worked it up – but found the metal of a horrid figure about the edges – nor c'd I polish it up. went again even'g. but c'd not make it much better tho' I worked long.[33]

By the middle of 1827 he was working on a telescope stand:

> Mr. Bishop had objected to legs, & the bottom piece of stand, so we were to have others. Went to White again evening – saw him fit the tenon, & then new underpieces ... 1827 May 2. Went to call on Mr Bishop – not at home; nor Bonnor ... then to Half St. again. at home: consulted w. him [Mr. Bishop] as to all particulars about stand, & he devised up a doz: plans, very ingenious, but too refined, but at last hit upon the very thing – shewed speculum to him, & after to Bonnor. hunted at Boughton's for hinges for tops of legs, according to Mr. B's suggestion – but none that w'd do ... After dinner to White – saw piece to be mortised in, nearly done – he advised me to let him rivet over the brass. so I was to get the brass made by another piece, wh. I took to Bonnor – not at home. Went to Edwards, got a very fine brass hinge to try – took it to White's – consulted then with him, & Jem Howlett for a long time – thought the hinge w'd be too large, & the screws w'd pull the joints about – besides the great size the legs w'd require. Jem bethought him of some old hinges he had – so I resolved to see them tomorrow & took back Edward's hinge. Jem had taken it to 2 other places in vain.[34]

Astronomy and scientific enquiry were not entirely lacking, even in the coldest of weather. He includes in his diary of 1826: 16 January, 'Experimenting with thermometer'; 17 January, 'Heard of new comet.

Too cloudy'; 18 January, 'Most celestial night – hunting comet in vain'; and on 28 January, 'Brilliant starlight. Grand zodiacal light. Fancied comet in Argo navis.' His observations were very regular at Tretire once his own Newtonian telescope was in working order.

Oxford

At 7 by Retaliator to Oxford arriving at a ¼ before 1 ... only a plain man inside.[35]

From the summer of 1826, when he went up to Oxford for three weeks, Thomas spent the first two terms of every year at the university until the summer of 1830, when he took his BA, returning two years later to take his MA. He was never in residence in the autumn term, presumably because it was convenient for him to be at Tretire during harvest, or perhaps because he preferred to have some time at home, especially during the long clear nights of autumn when observations would be productive.[36]

Thomas did not follow his father to Wadham College, but went instead to Magdalen Hall – one of the smaller, less prestigious colleges. Wadham was a 'hard-reading' serious college which tended towards the evangelical stance – and to that end it had no organ in the chapel until about 1860.[37] These reasons may have been enough for John to suggest a different college. He may also have felt that his son was not sufficiently interested in classics to become a first-class academic, and therefore a less expensive college would be adequate for him to pursue his more varied interests. When Thomas went up, Magdalen Hall (now Hertford College) had a popular and reforming Principal in Dr John McBride.[38] By 1850 the college was the fourth largest society in Oxford when there were ninety men in hall.

In the early nineteenth century, Oxford University was beginning to be transformed from a mere finishing school for the landed gentry to a place of serious study. When Thomas entered college examinations were a novelty, and although the main examinations at the end of the first and the third years were mainly oral, some written papers had been introduced. Classics – especially Greek – were the essential subjects, followed by mathematics. It was stressed that undergraduates need not confine their studies to these disciplines, and there were voluntary classes and lectures in other subjects, including astronomy. The university was still dominated by the clergy, and no-one could qualify for a degree without knowledge of the Gospels in Greek, the Thirty-Nine articles, and Joseph Butler's *Analogy of Religion* (1736).[39] Thomas no

Magdalen Hall, by J. Fisher, 1827.

doubt took this state of affairs for granted. He was conservative in his opinions, and would not have deplored the lack of liberal debate.

Thomas first took the coach to Oxford on 7 March 1826 in order to matriculate, and 'entered without the least fuss or bother'. The following morning he paid his fees to Dr McBride and, after visiting the Bodleian Library and the Radcliffe Observatory, returned to Gloucester. He was enrolled as a Gentleman Commoner, being the son of a gentleman and cadet of a family that held the rank of baronet. By 1826 the privileges awarded to this title had been eroded, but the rank still provided a certain status, and fees were more substantial. He returned on 13 June, and was allocated a room in college. Having bought 'a brown jug, got other things arranged' and acquired his scout[40] – a man named Joy – he went to dinner. He quickly adopted a pattern of life that was to continue more or less throughout his university period. Days at college were relaxed:

> Breakfasted with Wright, Rawlings, Tomes & Stansbury. Logic w. Tomes. Walked to Somerton w. Tomes, Giles, Kendall. Tomes & Rawlings to wine w. me & the latter 3 to tea. Tipsy Turnbull came in to bother us & complain of Cary's ripping up his gown.[41]

No doubt, being an only child Thomas was happy when on his own. However, he quietly made a number of friends and seemed very easy in the company of his peers. His friends – almost all from his own college – were the sons of 'gentlemen' and clergymen; and of these, two thirds were to become parsons.[42] He gradually added to his acquaintances, and rarely dropped friends that he had made. Indeed, he took care of them, giving one medicine for a minor illness, and visiting another frequently when he was sick. Robert Tomes, from Long Marston in Gloucestershire, became perhaps his best friend from the summer of 1827.

In the mornings, undergraduates held breakfast parties after chapel and attended a compulsory lecture in college. Their pastimes were innocent. Walking was the most popular form of exercise, and Thomas and a friend would stroll for two or three miles in the afternoon, often round Christ Church meadows. Thomas took no active part in sport, although he attended the boat races in 1827 and occasionally watched cricket at Cowley. On Sundays he went to church, usually with friends, often attending New College chapel, or St Mary's church. It seems to have been serious entertainment – the equivalent of a combined concert and lecture. One Sunday he 'Went w. Grimmett to New Coll. most beautiful windows. sticky sermon – but famous anthem "In the beginning" done exceedingly well'. The sermons were not always 'sticky' or 'stupid'. Milman 'preached finely for 1hr & a quarter, till he was hoarse'. Thomas delighted in the organ, and was pleased to get an introduction to Bennett, the organist at New College, who gave him piano lessons and allowed him to try the college instrument. In his second year he hired a piano for his rooms.

Attendance at the college chapel was less noteworthy. It was officially compulsory but was not always adhered to. Thomas must have been dutiful for at the end of several terms the Principal thanked him for his regular appearances. Dinner in hall was held at about five o'clock. Thomas took his turn to carve at High table (as a Gentleman Commoner this was probably where he sat), and there was the usual student banter about the food. One Shrove Tuesday a student compared the pancakes to 'eating Leather Breeches – they being also burnt to the bottom, as if the owner had stood too near the fire'.[43]

In the evenings the students took tea and wine together. Thomas held an occasional wine party – one of which was improved by a parcel from his father, containing cake, tongue, and 'beautiful bottles of Constantia – most delightful stuff'. They had 'famous discussions' about 'life, death, the universe and everything.'[44] Thomas briefly reported serious topics, but gave more diary space to ghost stories and anecdotes. There is no evidence of intimate talk, and no discussion of sex – racy or otherwise. But how could there be? Thomas's sexual awakening was slow and very gentle, and some of his friends may have been similar in their development. On only one occasion does he

The Sheldonian Theatre in the mid-nineteenth century.

comment that one of his friends had brought a 'very pretty girl' to a concert, and he noted the 'pretty eyes' of a girl on the coach to Gloucester.

Drunkenness with resultant violence, however, was common amongst some students, and Thomas was often kept awake by 'Horrible noise at night'. He was not censorious about alcohol, but deplored its ill-mannered effects and was cross when punishment seemed inadequate. When he heard that two students went drunk into hall and threw everything about he commented 'and they were rusticated 1 term only!!!'.

Thomas's time was also taken up in a most unusual manner. Excepting one-and-a-half terms, Mrs Webb lived in lodgings in Oxford while Thomas was at college. There is no documentary evidence to tell us why she was there; but considering how much time Thomas spent with her at Tretire and in Gloucester it can be assumed that it was better for her to be where her son could continue to entertain and amuse her. That it was rare for a mother to accompany her son can be adjudged by a sarcastic remark made in a letter of 1827 to John Webb from a friend: 'I suppose when Mrs. Webb has taken her Degree at Oxford she will rejoin you.'[45] Whether John Webb was kind to foist the occupation of his wife upon his dutiful son can be debated, but Thomas seems largely to have taken the imposition in his stride. He settled her into her lodgings, took her for walks, and persuaded her to attend services and concerts. He even took his friends to take tea with her. She had to be considered when events such as Commemoration of Benefactors loomed, for ladies needed tickets and he was assiduous in finding them for her. He was enthusiastic about his first Commemoration in June 1827, and took his mother to one of the concerts – a 'dreadful crush. full to suffocation' but 'Palestrina most glorious & delightful – capital seat – & everything agreeable'.

Another event which excited Thomas was the electoral contest between Peel and Inglis during the furore over Catholic Emancipation. Oxford had elected Robert Peel to Parliament as the anti-Catholic champion, but in 1829 Peel was forced to support the Emancipation Bill as it went through parliament. This horrified many at Oxford, including Thomas, who had cheered when the bill was first thrown out in 1828. He followed the election with enthusiasm, going each day to ascertain the progress of the vote. On the final day, when Members from the country came up to vote, Tomes and Thomas decided to disguise themselves and creep into the Sheldonian Theatre:

> [I] lent him spectacles & went & got hair cut ... I put on black & spectacles & took umbrella & made a horrid face & pushed in ... stayed from ¼ before 12 till 2 ... Peel 59, Inglis 79. Immense acclamation – followed by one cheer more ... Peel – immense hissing. Lord Eldon – an amazing cheer. The Pope – roars of laughter ... quite delighted. the house amazingly full, the entrance having been forced. A most glorious triumph: report of Peel's resignation.[46]

A week later the bill was passed by 'a shocking majority', but the campaigners did not give up. Thomas eagerly helped to collect names on a petition, and for the next fortnight did little else, obtaining 'the signatures of men of all sorts & sizes' from his own and other colleges. Like his father, Thomas was a Tory and no reformer. During the next two years he noted with disapprobation the moves towards parliamentary reform. 'BILL THROWN OUT! Well done!' was his comment in 1831.

There was the small matter of work. The daily college lecture – usually text-centred and more like a class – was attended by about fifteen students, often of widely varying abilities. Thomas had a liking and warm respect for the Principal, but he found his divinity lectures were long, often tiresome, and occasionally 'most grievous'. During one term he agreed to attend a course of his lectures only because there were very few of them, and writing up notes 'hurt' him very much. He did not attend many Latin lectures, taught by William James, the Vice-Principal, but the latter commented that Thomas's

Latin was already very good. There seem to have been no tutorials except before examinations, and most of Thomas's work was done in private study. Unlike many undergraduates Thomas did not pay for a private tutor to supplement the teaching in college. He listed Sophocles, Virgil, Livy, Euclid and Herodotus amongst his set texts, and he construed St Paul's epistles in Greek. At the beginning of 1829 he was urged to try for a class in mathematics, and he studied algebra, trigonometry, logic, optics, hydraulics and mechanics.

There were also lectures on other subjects, such as astronomy, at which attendance was entirely voluntary. It is known that Thomas attended the geology and palaeontology lectures of the flamboyant and eccentric William Buckland.[47]

His real interest, however, continued to be astronomy. He borrowed astronomical works from the college library, and read in the Bodleian Library – then not at all crowded. The Principal, aware of his enthusi-

The Radcliffe Observatory, Oxford, in 1814.

asm, took him to the Radcliffe Observatory in his first term, which greatly pleased Thomas:

> Most admirable invention of the Professors. Dollond 10 x 3½ feet – former upon most simple stand. Herschel 10ft. cost 40 or 50 guineas, 2 quadrants; Equatorial in garden. Zenith section & instrument. Offered to give me leave to Observe!![48]

He did not take much advantage of this offer, but when he was introduced to the new Savilian Professor of Astronomy, Stephen Rigaud, in 1827, he regularly attended his lectures, often called on him at his home to discuss details of mutual interest, and borrowed his books. Perhaps the state of the observatory was not all that he wished:

> Rare fun at that lecture for a Professor took us to see the Sun's spots with 3½ inch Dollond on the glass using a green & purple handglass but clouds continually made it too bright or too dark. Rigaud very kind & funny. I looked at it without a glass to his surprize as he said he thought the luminary of my eye must be 'boiled'. Went to see Equatorial. There we found 2 birds' nests between the windows & outside blinds. – eggs in one & birds in the other actually lying against the glass.[49]

Thomas's only other recorded visit to an observatory during his Oxford years was a guided tour of Greenwich which John Webb arranged during the short summer vacation of 1828. They were shown Halley's historic telescopes and other instruments.

Thomas took his final examination – a mixture of oral and written papers – in the summer of 1829. He had to take examinations in classics, divinity and mathematics, including some questions on astronomy. It was all 'very stiff work'. The differential calculus was 'evidently hopeless', though the hydraulics succeeded better than he hoped. After the mathematics *viva voce* the examiner asked Thomas what class he intended to obtain. 'Anything you w'd be pleased to give me,' he answered humbly. Two weeks later he was told that he had obtained second-class honours. In later years he was to say that he had been very idle, spending most of his time in desultory reading in the libraries;[50] but considering that in 1827 only 160 students took degrees, and of those 116 were in the lowest class, he need not have been so self-effacing.[51]

He spent two terms more in 1830 to fulfil his residency, using them to study more divinity and astronomy. On 2 June he took his degree, though it does not seem to have been a significant occasion, for all he wrote was: 'Then went & was made BA. very hot and fatiguing. Ma saw it'. However, he and his friends celebrated the event the previous evening with a 'Most excessively amusing party – lots of fun – slapping one another's thighs all around the table – Garwood's hunt story. Plenty of speechifying & amazing uproar for a steady party'.

In 1832 he went up – and mother went too – to take his Master's degree. This visit coincided with the second meeting of the British Association for the Advancement of Science.[52] It was said that a 'visit of the Association was not unlike the coming of a circus'. Local intellectuals had an opportunity to mingle with the eminent who came in droves. There was lavish entertainment, lectures by well-known speakers, and sometimes great arguments.

Dr McBride encouraged Thomas to join, and he became a life member of the Association. He attended every day. Excited to see and hear so many eminent scientists, he went to the chemistry, meteorology and geology sections, heard Airy[53] on comets and Willis on musical sounds, among other lectures and debates, and saw 'many beautiful experiments'. He went eagerly to the Ashmolean Museum to hear Faraday's guest lecture on electromagnetism, and spoke to Willis and Dalton[54] about 'thunder and musical tone'. Then, after a while:

> Faraday came & brought the apparatus thus, of which much was concealed by a linen wrapper, but what I could make out was thus: a – a round bar of iron 1 inch diam – b – a handle in which was wood and linen for convenience, but the essential part was a coil of copper wire of which the ends, each of 3 or 4 wires, some straight, some spiral, appeared at c. c. & were brought to meet at d. & e., by their elasticity. d. was a blunt point. e. a little tray containing mercury – but this was only, as someone said, to reflect the spark. d. & e. were so lightly in contact that thro' the elasticity of their arms which carried them, the least jar would make them separate. When therefore the two ends of a wire were laid on the magnet's poles it became itself a powerful magnet, & the copper wire had an electrical current passing through it – but unperceived – but the moment one end of a. was pulled from the magnet a small but very bright spark was seen between d. & e. & again when the contact was made. But all depended on the rapidity of the contact & separation. Tho' a bright day the sparks could be very plainly seen, & some were so bright they might well have been seen the length of the room. Many persons had the shock thro' their tongue & on gums, to see the flash – some did, some did not. I tried the gums – felt a shock, tho' doubtful whether the jar or the contact. But I am pretty sure it was the real shock.

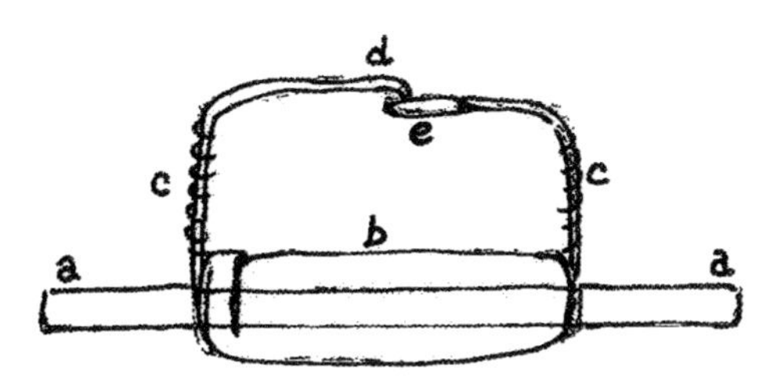

Thomas was also at the Sheldonian Theatre when William Smith – the 'father of Geology' and the 'beau ideal of John Bull' – was awarded a Gold Medal and Michael Faraday received an honorary degree. His notes about the meeting are the longest entries in the extant diaries.[55]

In those few weeks he saw all those at Oxford that he had valued – such as Stephen Rigaud and the Principal – and he attended sermons by his favourite preachers. Several of his friends were in residence at that time, and he parted from them, commenting regretfully that it was quite likely that he would never see them again. Then it was 'very hard fag packing' and cording boxes, organising mother – who may have been relieved to finish with all the academic debate to which she had been dragged – and off to the Regulator coach, where they could only find outside seats for their journey back to Gloucester.

Notes and references

1 T.W. Webb, in J. Webb and T.W. Webb, 'Anecdotes', MS in Hereford City Library.
2 W. Kirby and W. Spence, *An Introduction to Entomology*, 4 vols. (Longman, Rees, Orme, Brown & Green, London, 1815.) It became a popular handbook for the many beetle-hunters in early nineteenth-century Britain.
3 T.W. Webb, MSS in Hereford City Library. The eight extant notebooks of these observations cover the period 1817–61. They are thin soft-covered books, some with marbled covers, eight in number, seven of them 5 x 3 inches and the other 6 x 4 inches.
4 Ibid.
5 Ibid.
6 T.W. Webb, Preface to John Webb's *Memorials of the Civil War in Herefordshire*, T.W. Webb (ed.), 2 vols. (Longman, Green & Co, London, 1879).
7 Sir Henry Webb, seventh baronet, died in lodgings (and, one suspects, in the arms of his German mistress) in Esslingen, Germany, leaving effects of less than £5,000. No money came to Thomas.
8 Thanks are due to Adrian Harvey for information on Webb's ancestry. See also L.G. Pine (ed.), *Genealogical and Heraldic History of the Landed Gentry*, seventeenth edition (Burke's Peerage, London, 1952).
9 W.W. Webb (revised B. Frith), 'John Webb', *Oxford Dictionary of National Biography* (Oxford University Press, 2004).
10 Ibid.
11 Will of Thomas Webb of Hoxton, MS in Hereford Record Office.
12 W.W. Webb (n. 9).
13 See Chapter 5.
14 The situation may have been similar to that in Disraeli's novel *Vivian Grey* (1826), in which the father 'was for Eton but his lady one of those women whom nothing in the world can persuade that a public school is anything

but a place where boys are roasted alive, and so with tears, taunts and supplications, the point ... was conceded.'

15 A.C. Ranyard, obituary of T.W. Webb, *Monthly Notices of the Royal Astronomical Society*, **46** (4 February 1886), 198–201.
16 Announced in his diary (n. 27).
17 T.W. Webb, 'Meteorological Journal, 1825–26', MS in Hereford City Library.
18 T.W. Webb, Diary 1826–40, MS in Hereford City Library – hereafter referred to as Diary.
19 There is now a manuscript transcript of the diary in Hereford City Library.
20 Charles Dickens, *Pickwick Papers* (London, 1837). Webb's diary may be compared with that of W.E. Gladstone – who was almost a contemporary – for its similarity in style.
21 In his diary he once commented that they had mutton for dinner at home, adding grudgingly, 'but it was very sweet'. Goose is mentioned, otherwise nothing. Nor does he describe appearances.
22 Diary, 18 November 1826.
23 Diary, 16 August 1828.
24 Diary, 29 December 1829.
25 See Philip Barrett, *Barchester: English Cathedral Life in the Nineteenth Century* (SPCK Books, London, 1993).
26 Diary, 4 May 1826.
27 Diary, 20 July 1826.
28 The oldest music festival in the country started in 1724, and is still held each year at Gloucester, Hereford and Worcester in turn.
29 T.W. Webb, 'Observations on the Moon, 1826–55', MS in Hereford City Library.
30 Gell and Bradshaw, *Directory of Gloucester* (1820).
31 Diary, 5 April 1827.
32 See Chapter 8.
33 Diary, 28 September 1829.
34 Diary, 3 May 1827.
35 Diary, 11 November 1827.
36 This arrangement was permissible, but undergraduates had to be in residence for twelve terms. The academic year was divided into four terms: Hilary, Easter and Trinity (the summer term being split in two), and Michaelmas.
37 I am indebted to Dr Allan Chapman for this information.
38 John McBride (*b.*1778), Principal of Magdalen Hall/Hertford College, 1814–68. (*Dictionary of National Biography*, College History series, 1908, by Sidney Hamilton, Librarian, Hertford College.)
39 For much information in this chapter I am grateful to Michael Curthoys. See M.G. Brock and M.C. Curthoys, *History of the University of Oxford Vol.VI: Nineteenth Century Oxford, Pt.1* (Clarendon Press, Oxford, 1997).
40 Scout: a college servant.
41 Diary, 14 March 1827. Webb used only initials for the names of each of his

friends, but for clarity they have here been written in full.

42 In 1830 about half the students planned to be parsons and one third were parsons' sons.

43 Diary, 3 March 1829.

44 Douglas Adams, *The Hitch-hiker's Guide to the Galaxy* (Ballantine Reissue Edition, London, 1995).

45 Mrs Westfalling of Rudhall, MS letters in Hereford Record Office. The mother of John Ruskin also accompanied her son to Oxford, but she wished to watch over his studies.

46 Diary, 28 February 1829.

47 William Buckland (1784–1856): Reader (1827); Professor of Mineralogy and then Professor of Geology; Fellow of the Royal Society; President of the British Association for the Advancement of Science (BAAS), 1831; President of the Geological Society, 1840; Dean of Westminster, 1840; see *Oxford Dictionary of National Biography* (Oxford University Press, 2004).

48 Diary, 4 July 1826.

49 Diary, 30 May 1828.

50 A.C. Ranyard, obituary of Webb (n. 15).

51 *Oxford Journal*, July 1827.

52 The BAAS was founded in 1831; see David Knight: *The Age of Science* (Blackwell, Oxford, 1986).

53 Sir George Biddell Airy (1801–1892); see Allan Chapman, *The Victorian Amateur Astronomer: Independent Astronomical Research in Britain, 1820–1920*, Chapter 1 (Wiley–Praxis, Chichester and New York, 1998).

54 John Dalton (1766–1844) was a pioneer of chemical science. Robert Willis (1800–1875) was a Cambridge professor with an international reputation as a mechanical engineer, and was also a talented musician.

55 Diary, 19 May to 6 June 1832.

Chapter 2

Courtship, curacies and marriage, 1826–56

Janet H. Robinson

The portrait of Thomas Webb in the frontispiece of the fifth edition of *Celestial Objects for Common Telescopes* suggests a kind and avuncular Victorian clergyman. This picture is reinforced by reading his letters of advice in such periodicals as the *English Mechanic*, and is confirmed by the laudatory words of those who wrote his obituaries. They depict him as settled and stable in his maturity, with a gentle wit and playful humour but with little hesitancy or shyness in his manner. It was not always so . . .

Courtship

> *I also state that all the shorthand cyphers etc. in the Journal have related to what I have seen of this young lady as I took great interest in her.*[1]

In 1826, when Thomas was nineteen years old, a form of shorthand appeared in his diary, and its incidence occurred more frequently in the years up to 1830.[2] Transcribed (the shorthand is here represented in italics), it shows that he believed that anything he wrote about young ladies needed encoding, even though it revealed nothing intimate: 'Mrs Webb said: *I think Susan [Harvey] is a most likeable lady*.' Some neighbourly gossip about young women was also written in shorthand – such as the following diary entry after a visit to Ross Flower Show on 9 October 1828:

> Miss Holder said *Somebody had been so ill-natured as to say Miss Harvey did it out of display* (Shame!) *and therefore she would not be likely to play again, but*

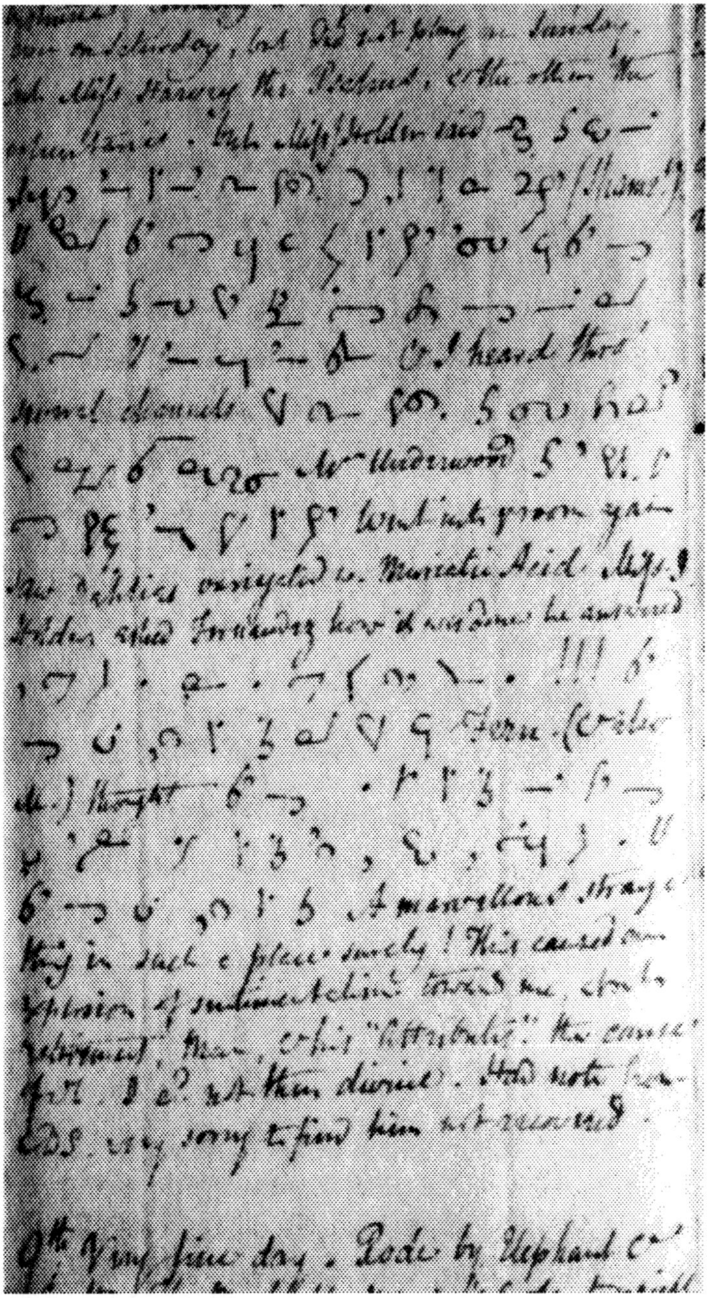

A page of Webb's diary, showing his shorthand.

> *she said nobody who had seen the lads would have said so for they were all as white as sheets* & I heard thro' several channels *that Miss Harvey had gone home from the flower shew unhappy*. Mr. Underwood *had a party & would probably ask her to play*. Went into the room again. saw Dahlias variegated w. Muriatic acid. Miss J Holder asked Fernandez how it was done. He answered *I will tell you if you will let me kiss you !!! She said No, I am too old for that but* Fern (& also M.) thought *she said you are too old, so he said in a huff: Oh too old am I, then I won't tell you & she said, No, I'm too old*. A marvellous strange thing in such a place surely!

Thomas's mother wanted him to take an interest in young ladies, and made heavy hints, 'betting' that Susan would make a nice wife, or remarking that he would do well to see a little more of Fanny Aveline. Her son reported her remarks but showed little enthusiasm for the ladies concerned.

In 1829, however, he met Helen Duff in Gloucester. She was a niece of Dr John Baron, a longstanding friend of John Webb. Baron was an eminent physician in Gloucester, and had written the first biography of Edward Jenner, the index of which Thomas spent some time checking. Mrs Webb was quick to scent matchmaking, and commented on Miss Duff's elegance and 'ladylike' qualities. This time her son agreed with her. He began writing verse, and read poetry which moved him to tears, even quoting 'Give me but my Arab Steed' adding in shorthand ' – *and Miss Duff*'.[3]

He would write casually that Miss Baron and Miss Duff had called, and at the end of the day's entry there would be two small circles, thus: OO. Subsequently, other entries had similar circles, sometimes larger, or one or more crosses: +++. The young people walked and talked together – though probably never entirely alone. Miss Duff seems to have been of a flirtatious nature, for Thomas reported, no doubt with a frisson, that she had '*threatened*' to put him in the canal and leave him there, and '*threatened*' she would object to the Bishop about his ordination.[4] The circles and crosses in the diary grow more agitated, and on 27 July 1830 the shorthand ceased, never to reappear. A long, sad entry in the diary recorded that the girl was returning to Scotland.

Thomas explained that the OO indicated that he had seen or spoken with her and the +++ signified that he had 'crosses or misgivings'. He added ' I do not know that this diary may ever fall into any hands but my own – & yet I think it not impossible since I seldom destroy papers' – indicating that while he was not seeking public record he was prepared for the eventuality. He confessed that he had for her 'a most sincere and pure affection & esteem. & had it pleased God, should have been, I fear, too happy had she returned it'.

There was a last meeting when he and Helen found the courage to profess that they had certain feelings beyond mere friendship towards each other. Thomas, taken by surprise and aware that he had not asked for the consent of his parents, felt he could go no further, and was forced to say, 'It is a sore subject.' She said, 'It is a sore subject and we'll drop it.' Thomas added in his diary account that evening: 'It was quite evident & very delightful, tho' very embarrassing to me to find she was so predisposed.' He spoke to his parents, and it was agreed 'to leave it to a good providence. The uncle to make the first move'. Like many a young man of the period he made no further efforts to promote the match. He hoped, when he visited the Barons, that they would refer to his future prospects; but when they said nothing he grew increasingly disappointed. By May 1831 he seemed to have come to terms with his feelings because he was surprised and almost vexed to find that he was so happy, feeling that as 'an absent lover' he should not have been so. In July he coped very well with the first anniversary of their parting, and two days later he was more interested in the performance of his telescope, which showed ζ Aquarii beautifully. It was an assay into romance that was not to be repeated for some time.

Curacies

With very deep self-abasement & the most concentrated intensity of feeling, I was ordained Priest. Thanks be to God.[5]

In July 1830 Thomas was ordained deacon at Hereford Cathedral, and was priested there in August of the following year. He began his clerical career as curate of Pencoed – a small, plain church just over a mile from home – whilst also assisting his father at Tretire. In November 1830 the family returned permanently to Tretire, and once there, besides the practice of his profession, Thomas's accustomed habits and interests of country life were resumed. He tended the glebe and garden, and walked and played cards with his mother.

Life was not always calm and peaceful. The propertied classes in England, shaken to the foundations by the French Revolution, saw many signs of political unrest throughout the early years of the nineteenth century. In his diary in 1831 Thomas mentions riots in Derby and South Wales. These riots were probably activated by discontent over wages and the ferment of parliamentary reform. Such protests were not serious in Herefordshire, but there was enough disquiet for property owners to take some precautions and see even theft as a

St Weonards Church, Herefordshire.

portent of something worse. John Webb issued a Public Warning after some trespass at Tretire, putting notices around the villages, and Thomas recorded his wariness of strange noises at night, once creeping round the house with loaded pistols, cutting 'a most absurd figure with a rattle & pistol in one hand, a lighted candle in the other, & a night-cap upon my head'.[6]

He continued to live at home for the next fourteen years. His curacy of Pencoed was terminated in 1833 when the vicar took residence. Fortunately, Sir Hungerford Hoskyns, of the nearby Harewood estate, asked him to be his chaplain, and he looked after Harewood church and the people of the estate until 1834, when he was offered the curacy of St Weonards, a larger church and parish on the road to Monmouth and two miles from Tretire.

In 1833 Thomas's astronomical studies were augmented when his father announced that he had ordered for him

> . . . a 5ft Achromatic $3\frac{1}{2}$ in. aperture from Tulley. God grant I may use it to his glory – A day to be remembered . . . It is to be a 100 guinea Telescope. This is handsome indeed! – but he says I cost him so little at Oxford he had determined to make me a present.[7]

Thomas continued his observations, detailing them in notebooks dedicated to the purpose. In September 1835 he was asked by the editor of *The Analyst* to write an article. Clearly excited, he set about it, finished it in less than two days and sent it off. 'Hints to Observers of Halley's Comet' (which was then visible) was probably his first published article, and was followed two months later by 'On the influence of Comets' in the same journal. This was a strong contentious article quite unlike the gentle instructional works of his later years. An abstract of his paper on 'Lunar Volcanoes' was read at the 1838 meeting of the British Association for the Advancement of Science.[8] He also read a lecture on comets at a soirée in Hereford at the beginning of 1840, and later gave a talk on meteor showers to the Gloucester Literary and Scientific Association. About this time he began to give lectures in Cheltenham, and continued doing so for many years. He may have lectured to several associations, but it is documented that he spoke to the Working Man's Club[9] and, from 1865, to the girls of Cheltenham Ladies' College.

Gradually other activities crept in which, although they were often part of his position and way of life, reveal the cast of Thomas's mind and character. Many parsons in those days took some responsibility for the physical health of their parishioners, even if it was only charity in the form of medical comforts such as oranges or wine, or taking 'a little brandy & water & paregoric to Netherton for Davis'; but Thomas followed the practice of his father and took a more active rôle. In 1834 he vaccinated children in the local parishes,[10] and he took detailed notes of any medical recipes which seemed to have been efficacious; for example, this prescription for ague (fever):

> 2 drams apiece of Jesuit Bark, Snake root, & Salt of Wormwood mixed in Whey. To be made 3 doses of, 1 to be taken before the fit also to be made into an electuary in 9 parts, & taken for 9 mornings.[11]

A treatment in which he was very interested, and had first used on his mother in 1826, was a form of electrotherapy. Shocks were applied to the patient by the use of an electrostatic machine. In 1833 he persuaded a parishioner to try this treatment for her rheumatism. His diary entries suggest that he was more interested in the science than in the patient. When he tried a different machine in the form of Leyden jars it 'quite overpowered her & she was ready to faint & cried greatly – & the bumps upon the arm where the chair came in contact were surprizing as the shocks were very small'. Not surprisingly she did not come again. Two weeks later Thomas reported with honesty and possible chagrin: 'rode to see Mrs Preece. [She] had been cured!! apparently by Fletcher, with peroxide of iron. He maintained it was no rheumatism whatever! an affliction of the heart.'[12]

He had little leisure time, but occasionally he made day expeditions on his horse, such as that to the Vale of Ewyas in the Black Mountains. There he was 'greatly delighted' by the ruins of Llanthony Abbey, but, like the Revd Francis Kilvert[13] thirty-five years later, was annoyed by a 'vulgar, noisy party'. After a return journey of more than thirty miles, which shows a degree of stamina and fitness, he could say:

> As far as scenery was concerned, this was the most remarkable day I ever passed, nor had I the least idea of the wild magnificence of the Nant Honddu or the beauty of the Abbey.[14]

His one extended journey during these years was a business trip to buy an organ for St Weonards church; when he also stayed with relations and visited places of interest. He caught the Lion coach to Bristol and, having seen fireworks 'enough to singe the whiskers of a whole regiment of dragoons',[15] went on by coach to Salisbury, where he caught a train for the first time and went to London. Having dealt with the order for the organ at Walkers he used another train from New Cross to visit an asylum, and the following week bought a 25-shilling ticket from Euston to Coventry. There he was met by his uncle and taken to Leamington. He wrote to his parents:

> I am fairly delighted with train travel – not that I have yet made up my mind to one from Tretire to St Weonards – but as compared with the barbarous system of coaching it is quite beautiful, & all the arrangements approach very near perfection so that my mind has been quite altered about it.[16]

He had indeed changed his mind, for ten years previously he was equally delighted when there were objections to the Ross–Gloucester railway, and a project for a Tram Road in Cusop, near Hay (a project which came to nothing), greatly vexed him: 'One of the few, very few remaining wild and untrodden regions of the county – is this to be entirely spoiled?' He continued to use the railways for all his longer journeys to London and to the continent. Nonetheless, in 1876, when it was proposed to extend the Golden Valley Railway from Dorstone to Hay, the line to come close to the front of the vicarage, he told Arthur Ranyard:

> We are still in a worry, all the neighbourhood, about the projected Railway – & not hopeless as to its being checked, or at least postponed ... The property [his patron's estate] will be greatly damaged by the passage

> of the Railway through it – and it is to be opposed – or at any rate full compensation demanded.[17]

He wrote in protest to the parliamentary committee, claiming that it would decrease the value of the glebe land. He would not be the first nor last to cry 'not in my backyard'.

Marriage

A woman of very rare gifts and of a most generous heart.[18]

In 1843 Thomas Webb married Henrietta Montagu Wyatt. The extant diary stops in October 1840, so there remains no description of the meeting or the courtship like that which described the romance with Miss Duff. There is only surmise, conjecture and suggestion. Henrietta was a relatively local girl, so they may have met at local events in years past or at Wyesham, near Monmouth, the home of the Tudors. Thomas Tudor was a draughtsman and land surveyor who designed the case and loft for the new organ at St Weonards and had executed other professional work for John Webb. His wife was aunt to Henrietta Wyatt, and Webb's diary of 1840 records a number of visits to the Tudor household – visits that elicited enthusiastic comments such as 'delightful day at Wyesham. Glorious evening'. His notebooks of that period include one playfully entitled 'Observations on Natural Phenomena on behalf of the United Tretire and Wyesham Astronomical Association by Monasticus and Theodorus', containing entries from 1838 to 1843.[19] This appears to have been a way of enthusing Owen – Thomas Tudor's young son – in astronomy and science. Perhaps it also created opportunities to meet Henrietta, using Owen as the go-between. The entries are few in number and in the handwriting of Thomas and that of a child. The last entry is some ten days before the wedding, and is in a third hand. The writing is very similar to that of Henrietta as seen in her later letters.

Henrietta Wyatt was ten years younger than Thomas, and was born in 1816 at Troy House, Mitchel Troy, south of Monmouth. She was the youngest child of Arthur and Arabella Wyatt. The family had aristocratic connexions through their mother, and the Wyatt dynasty produced painters, inventors, land agents and surveyors, plus a host of architects.[20] Henrietta's father was land agent to the Duke of Beaufort – a position which her brother Oswald assumed upon the death of his father in 1833. Her mother died when she was two years old, but she had four sisters and two brothers who no doubt spoiled her, and the family was further augmented by their cousins Thomas Henry and Matthew Digby Wyatt, who spent their summer holidays at Troy House. The house was large, late seventeenth century, and 'wonderfully undisturbed, full of old furniture and redolent of faded grandeur'.[21] It was a lively and intelligent band of children who romped amidst the vast gardens, sketched the ruins at Tintern and Raglan, and took drawing lessons with David Cox. It was a very different childhood from that experienced by Thomas.

We know that by the beginning of 1843 Thomas's courtship of Henrietta was a known and settled fact, for John Webb's pocket diary for 1843 is in the public domain.[22] In it we read that Thomas went to Monmouth about once a week, and each month of that spring stayed for two or three days at Troy House. In March his father gave him 17s 6½d for shirting – no doubt for Thomas's bottom drawer.

On 5 May Thomas's grandmother and aunt arrived from London. Thomas plied between Tretire and Troy, and went to Abergavenny for a marriage licence. He asked his father to take his duties at St Weonards or get another clergyman at 1 guinea for Sunday duty while he was away, and marched John to the church to ensure that his father 'took his directions' about his beloved organ. On 15 May John wrote laconically: 'TWW to Troy. The marriage to be tomorrow'. Mrs Webb had been 'taken ill' on 24 April, and may have been an embarrassment at Troy. Random evidence in family letters suggests that from 1839 she had developed a form of Alzheimer's disease. Whatever the reason, she did not attend the wedding. John Webb took his sister 'in a Fly to Mitchel Troy church to be present at the wedding. 6 carriages – Quod faciente ut felix: te Deus ora. Flyman 7s.6d.' Arthur – Henrietta's elder brother, by then a curate – solemnised the marriage. Thomas and Henrietta then went on honeymoon for almost a month. It is not known where they went, but John gave his son £50 for his expenses, of which he dutifully returned £4 1s 11d. On 8 June 'Mr Wyatt's man brought over Mrs T.W.W.'s Boxes & bath'. Two days later the couple returned, and the new Mrs Webb went to church at Tretire the following day for the first time. They then made formal visits to houses in the district, when Henrietta, in accordance with custom, no doubt wore her wedding dress at dinner.

They lived at the rectory, but made frequent visits to Troy House throughout the rest of the year, and sometimes stayed there for four or five days at a time. Henrietta must have missed her brothers and sisters, and life at Tretire may have been difficult and rather dull, compared with a more lively social life in her previous home.

In 1844 Thomas took up posts in Gloucester Cathedral, being precentor, librarian and a minor canon. He and Henrietta moved to a house in the Cloisters at Gloucester, taking delivery on 1 August of a second-hand organ in a mahogany case.[23] Although Thomas continued to take services in his father's parish when the latter was away, and often visited Tretire, the couple must have enjoyed a more independent life together.

An indefatigable recorder, he kept a hard-backed marbled-paper-covered notebook[24] during this period, and filled it with notes, anecdotes, quotations and newspaper cuttings on a variety of topics. On one page there are notes on Cassegrain reflectors, moles, crickets, toads, achromatic telescopes, and the air. He also continued to look for and record antiquities. He took advantage of the better communications from Gloucester to visit London, attending various lectures, and in 1848 he was present at the annual visitation of the Board of Visitors to the Royal Observatory, Greenwich. Both he and Henrietta attended the annual meeting of the BAAS at Swansea, where he attended lectures on metallurgical operations, and went on to Tenby in Pembrokeshire for a holiday. In November of the same year, Thomas took a practical approach to the popularisation of astronomy by moving his Tulley refractor, complete with deep blue solar caps, into a public street so that at least eighty or ninety people could observe the transit of Mercury.[25]

In July 1849 Thomas's mother died. Her sister wrote 'her happy spirit [h]as had a merciful deliverance' – and indeed her poor mental health must have made life difficult for all the family. In his obituary of Thomas, Arthur Ranyard – a close friend of the Webbs – stated that Mrs Sarah Webb had died when Thomas was a small child.[26] Thomas had always been a devoted son to his mother, but perhaps he did not wish to talk about her, remembering later years of mental illness and frailty. He may have been ashamed, or perhaps the memories were too painful. This may account for Ranyard's error – one which was copied by other writers, including T.H.E.C. Espin, another close friend and one of Webb's executors.[27]

Thomas had resigned his position at Gloucester in June 1849, and the couple may have returned to Tretire in order to support his parents. Conveniently, he was then appointed as curate at Ganarew, Monmouthshire, and he and Henrietta moved in to Rectory House in November and lived there for four years. Ganarew was nicely situated between Tretire and Mitchel Troy. At Troy House the nursery was rapidly filling with the children of Henrietta's brother, Oswald. There were four by 1850, and by 1867 Oswald and his wife, Louisa, had fourteen children. It is only surmise, but one suspects, with the evidence of their later interest in young people, that Henrietta and Thomas may have yearned over the nursery of the Wyatts, longing to have children of their own.

Thomas was elected as Fellow of the Royal Astronomical Society in 1850. At this time he began to borrow books from the library on a regular basis – such volumes as Struve's *Catalogue of Double Stars*, the Royal Society's *Philosophical Transactions*, and books by Schumacher, Beer and Mädler. He sent about two notes every year for publication in *Monthly Notices of the Royal Astronomical Society*, on such subjects as the zodiacal light, comets and eruptions on the Moon.

In 1854 he was officially licensed as curate to his father – a job he had been fulfilling unofficially for many years. Resigning his position at Ganarew, he and Henrietta moved into the rectory at Tretire. John had bought the patronage of Tretire with Michaelchurch from Guy's Hospital some years before. In 1855, with the aid of Henrietta's architect cousin and brother-in-law, Thomas Henry Wyatt,[28] John undertook a massive restoration of the church, mainly at his own expense. At the same time, £450 was spent on modernising and extending the rectory. John was now seventy-eight years old, and no doubt he intended that his son should gradually take over the living.

In 1856 this plan was overturned when Frances Rigby Brodbelt Stallard Penoyre offered Thomas a living at Hardwicke, on the western border of Herefordshire. John Webb was a lifelong friend of the family who owned The Moor estate, with property and land situated in Hardwicke and the neighbouring parish of Clifford. It was a place that Thomas knew well; as a child and young man he had visited the Penoyres with his father. No doubt there was serious discussion in the rectory at Tretire when this offer was made. It was, of course, an opportunity for Thomas to improve his financial position, and Henrietta may have been eager to preside over her own household and enjoy some independence; but it placed John's plans for Tretire in jeopardy. Whatever the deliberations, the decision was quickly made to accept the offer, and Thomas and Henrietta prepared to move to the parsonage at Hardwicke.

Notes and references

1 Diary, 27 July 1830.

2 The shorthand was that of William Mavour: *Universal Stenography* (1780). The book ran into several editions in the early nineteenth century, with slight differences. Thomas used the first edition – possibly his father's copy.

3 Poem by Caroline Norton (1806–1877). Miss Duff and Webb's beloved horse, Pretty Polly, seem somewhat confused in his mind, for the poem goes on:

My beautiful, my beautiful! that standest meekly by,
With thy proudly arched and glossy neck, and dark and fiery eye!

Fret not to roam the desert now, with all thy winged speed:
I may not mount on thee again! – thou'rt sold, my Arab steed.

4 It was customary that before a man was ordained, people were allowed to object or make known any impediment to ordination to the priesthood.
5 Diary, 6 August 1831.
6 Diary, 26 April 1832.
7 Diary, 8 June 1833.
8 BAAS Annual Report, 1838.
9 See Chapter 15.
10 It was more than thirty-five years since vaccination (originated by Edward Jenner) against smallpox had become regular practice and a fashionable thing for gentry and clergy to do for their various dependants. An 1840 Act of Parliament forbade the practice of vaccination by lay people.
11 Diary, 31 March 1832. Jesuit bark: Peruvian bark, or cinchona – a specie of tree associated with quinine.
12 Diary, 14 March 1833.
13 See W. Plomer (ed.), *Kilvert's Diary: Selections from the Diary of the Rev. Francis. J. Kilvert, 1870-79* (Jonathan Cape, London, 1938–40).
14 Diary, 28 July 1835.
15 T.W. Webb, letter to his parents, 13 February 1839, MS in Hereford Record Office.
16 Ibid., 27 February 1840.
17 Webb to Ranyard, 21 December 1876.
18 Extract from the memorial plaque to Henrietta Webb in Holy Trinity church, Hardwicke.
19 MS in Hereford City Library. Unfortunately his diaries for 1838 and 1839 are missing – perhaps deliberately so.
20 See J.M. Robinson, *The Wyatts: An Architectural Dynasty* (Oxford University Press, 1979).
21 Ibid.
22 MS in Hereford Record Office. Kindly loaned by Tony Brown.
23 See Chapter 4, n. 37.
24 MS in Hereford Cathedral Library.
25 I am indebted to Alan Dowdell for this information.
26 Arthur C. Ranyard, Report to the AGM of the Royal Astronomical Society, February 1886.
27 T.H.E.C. Espin, 'T.W. Webb – A Reminiscence', in Webb's *Celestial Objects for Common Telescopes*, vol. 1 (fifth edition, revised by Espin, London, 1893), p. xv.
28 See Chapter 3, n. 22.

Chapter 3

The Hardwicke years, 1856–85

Mark Robinson and Janet H. Robinson

> *I am obsessed with heaps of letters and papers – but the silence and the climate and the garden of this place are delicious – Thank God for such a home.*[1]

> House-moving, which has been my late employment – and the labours of a large and new Parish – have taken up my time of late, and I do not look forward to much leisure.[2]

So Thomas Webb described his new parish soon after arriving to take up duties as the second vicar of the new church at Hardwicke.[3]

The parish lies now as then, on the fringes of Herefordshire, its western boundary only a mile from the border with Wales. There is a church, but no inns or shops, and the scattered dwellings have names such as Llanerchycoed, Archenfield, Broadmeadow and Penylan. There are about 250 people in the parish – much the same as in 1856, so hardly 'large' in terms of parishioners – and more than thirty small farms. In the nineteenth century there were at least two beer houses, a number of tiny shops, and tradesmen providing services to the local population. The rolling landscape contains hills that rise to over 300 metres and form almost a natural amphitheatre around the church that stands in the centre of the parish. Numerous sheep and cattle graze the small undulating, oak-studded fields. Mixed woodland patches the fields and steep slopes and clings to the sides of the dingles. These narrow little valleys channel water off the hills. In summer the water trickles down through weed and rock; in winter the brooks, thick with red mud, roar and race to the River Wye.

Two minor roads and many twisting lanes thread through the parish. Footpaths, often ill defined, link farms, or go by forgotten ways to church and the primary school that stands on the edge of the parish of Clifford. Even today the tranquility is torn apart only by the occasional thunder of low-flying aircraft. At night there is no light pollution, so the sky above Hardwicke can be brilliant with stars.

The church, designed in the Gothic style by Thomas Tudor, was consecrated in 1853. It was built ostensibly because the parish church at Clifford was considered inconveniently situated for many of the parishioners, and was too small. It was after the untimely death of the son-in-law of the patron and first incumbent that the living was offered to Thomas. The stone vicarage – also built through the generosity of Mrs Penoyre – was relatively small and plain, but was extended less than five years after the arrival of the Webbs to provide appropriate accommodation for John Webb in his retirement.

Hardwicke is two miles from Hay-on-Wye, otherwise known as Y Gelli. Hay is a small market town straddling the border of England and Wales. In the latter half of the nineteenth century it was usually called Hay or The Hay, and had a population of just under 2,000 – larger than at the beginning of the twenty-first century. In the Webbs' time there were more than thirty inns and beer houses. It would have been the first recourse for shopping for the inhabitants of the vicarage to supplement the food grown in the garden and raised on the glebe.

Beyond Hay and the river Wye lie the Black Mountains, the Brecon Beacons, Radnorshire and wild Wales. Twenty miles to the east of Hardwicke is Hereford, cathedral city and county town. When the Webbs came the railway had only just reached Hereford, and it was several years before direct journeys to London could be made by train. Hay and Hereford were linked by rail in 1864.

Thomas and Henrietta settled in Hardwicke with apparent content. As at Tretire they had an outside man and three indoor servants, plus part-time employees. Unlike Thomas's mother they seemed to have a warm relationship with their servants. James Willis, named in the censii as groom and then coachman, stayed for the whole time the Webbs were in Hardwicke, was with Thomas when the latter died and was left £150 in his will. One young woman came with them from the Tretire area and soon left them to marry a local man, but the Webbs were sufficiently fond of her to put up a plaque in the church when she died, presumably in childbirth.

Thomas's professional work at Hardwicke continued in the same pattern as his years as a curate, and parish visiting and the Sunday services were the main features.[4] Being a conscientious pastor his duties must have occupied much of his week, so, as Arthur Mee pointed out in

T.W. Webb. (From Arthur Mee's *Observational Astronomy*.)

Arthur Cowper Ranyard in 1879.

his obituary of Webb, Thomas had no 'abundance of leisure to devote to scientific pursuits'.[5] It is clear, however, from his extant notebooks and letters, that many nights were devoted to observation and his days were infused with scientific curiosity. On walks through the parish, or talking to servants, neighbours and parishioners, he would note and discuss weather, astronomical events and anything of natural interest. For example, on 4 March 1871 Thomas recorded that a parishioner had sighted a bright meteor. He questioned the man and together they compiled a clear account. 'The head was about $\frac{1}{3}$ as large as the Full Moon. It was surrounded by some sparks; it threw such a light upon the ground as to shew all the growing wheat in the field around him.' Thomas then drew a sketch which he said was a copy of that made and shown to the parishioner for verification.[6] The vicarage servants too either became genuinely informed and interested or, knowing it would please their employer, took sufficient notice of astronomical events to be able to answer pertinent questions.

'Stargazing' was made more comfortable when in 1866 John Webb purchased for his son a Newtonian reflector with a $9\frac{1}{3}$-inch silver-on-glass mirror by With and a Berthon equatorial mounting,[7] and an observatory. Up to that time all of Thomas's astronomical studies were pursued by means of a telescope set up in the garden. G.F. Chambers wrote:

> I shall never forget the feeling of blank astonishment which crept over my mind one day when (in, I think, the year 1864) Mr Webb told me that the first edition of his book, and all his magazine articles up to that date describing double stars and clusters, were founded on studies pursued by means of a telescope set up in his garden and not equatorially mounted.[8]

The observatory was of the 'Romsey' type, designed by the Revd E.L. Berthon.[9] This twelve-sided wood and canvas building with a conical rotating opening roof was situated in the front garden a few yards to the south-east of the house. Webb wrote:

> Weather and bother have conspired to reduce my Observatory experience within very narrow limits. But after all I find it such comfort as I little expected ever to possess. Instead of hauling in and out a long iron carronade weighing 50–60 lbs. – a process which in the utter uncertainty of English definition, often made me rather glad of an excuse for avoiding it, I can run over to the Observatory in any weather when there is a bit of blue, get a porthole open and the great gun in position in 3 m[inutes] – see what I want and have everything away tight in 1 m[inute] more and into the house again. This is charming.[10]

Before moving to Hardwicke Thomas had begun work on his most celebrated book, *Celestial Objects for Common Telescopes*, first published by Longmans in 1859,[11] and had also contributed to various scientific periodicals.[12] His many articles and letters – especially his contributions to the *English Mechanic*, which was first published in 1865 – involved him in voluminous correspondence with amateurs and beginners in astronomy. He also corresponded with astronomers abroad. Indeed, letters were so numerous that Webb had a private postbag delivered straight from Hereford – no doubt saving the Hay postmaster a great deal of work. Mee stated that Webb was 'A charming letter-writer [who] delighted beyond measure to assist and encourage the beginner, and the post was the bearer of countless epistles to the young stargazer fraught with wise counsel and kindly suggestions.'[13] Certainly many of his letters reveal a generous humour and a puckish wit.

One such youthful correspondent was Arthur C. Ranyard.[14] In May 1858 he wrote to Webb asking for advice as to what telescope he should buy, and Thomas replied extensively with no hint of condescension to the thirteen-year-old:

> Dear Sir, It always gives me especial pleasure to assist any young or inexperienced student, and therefore I was very glad to hear from yourself and beg that, if you feel so disposed, you would write to me at any time,

and tell me of any difficulty that may occur to you, though I am well aware that five minutes talk is worth many letters, but I am so seldom in London that I fear we have not much chance of meeting. Yours very faithfully, Thomas William Webb.[15]

Towards the end of September 1858 the second letter, of four quarto sides, gives advice on the care of the eyepiece, which presumably Ranyard had purchased. He suggested that the boy study the Great Comet then visible and advised him how to do so, including two little sketches of the tail. There were further exchanges of letters, and in August 1859 Thomas wrote a long letter which in tone could have been written to a close adult friend. He advised Ranyard on the purchase of a telescope, suggesting a Birmingham optician rather than one in London because the instrument would be more reasonable in price, 'or, if you do not care for appearances, I could get you one still cheaper, fitted up in a rough but effective way, by a very ingenious friend of mine in Hereford'.[16, 17] He continued:

A little work of mine, 'Celestial Objects for Common Telescopes' will be published by Longmans very shortly. I have had it in hand for a long while, having had so many things to attend to which took up unavoidably a great deal of time so that I could make but slow progress. At first I made it too long, so that my publisher thought the expense would be too great – so I had to abridge it – which occupied me six weeks – and then he considered that I had overdone it in point of brevity, and I had to expand it somewhat again. I am very glad it is at an end at last.[18]

Two months later he sent Ranyard eight tightly written quarto sides crammed with further information about buying a telescope – makers, cost, opinions of their quality. Addresses ranged from Hereford, Paris, Munich (Steinheil and Merz), Birmingham (Parker and Wray), London (Cooke, Dallmeyer and Bardou) and the USA (Clark). He commented on secondhand instruments advertised in the *Astronomical Register*, offered the use of his name as an introduction, discussed the merits of English versus German equatorial mounts, and concluded:

... but if your object is – like mine – merely what is called star gazing – the viewing of the wonders & glories of creation under ordinary circumstances, it might be worth your while to consider before finally deciding, the comparative merits of the silvered glass reflector. PS [written sideways at top left of first page] Thank you very much for your kind expressions about my little book. I am very glad it has given you so much pleasure.[19]

These letters, both frequent and detailed, are a valuable insight into the care and labour which he was prepared to expend upon fellow enthusiasts. Other correspondents have recorded their debt to Webb's industry. Both Arthur Mee and G.F. Chambers mention in their memoirs that they had kept many letters from Webb which they cherished. It is known that he wrote many answers to letters from many parts of the world, and in a letter to the British Astronomical Association in 1891, Mr Foulkes, a chaplain to the Forces, stated that Webb had written frequently to him – not only on astronomical subjects, but on such topics as 'how a clergyman, with no private means, might venture to take a living worth £250 a year'.

The correspondence with Ranyard continued, the two met in London, and it was not long before Ranyard was making visits to Hardwicke. That Arthur Ranyard valued the friendship of the couple – a friendship that lasted the whole of Webb's life – may be judged by the fact that he kept some three hundred letters sent to him by Thomas and those written by Henrietta – probably almost the complete correspondence. The requests to do small, and not so small, errands for the Webbs in London must sometimes have tried his patience, but the relationship was warm. Webb was apologetic about his demands on the younger man.

> My dearest Arthur, How sorry I am to be such a bore to you and to give such repeated proofs of intense stupidity. 'Tis a sad misfortune to have such a friend ... [20]

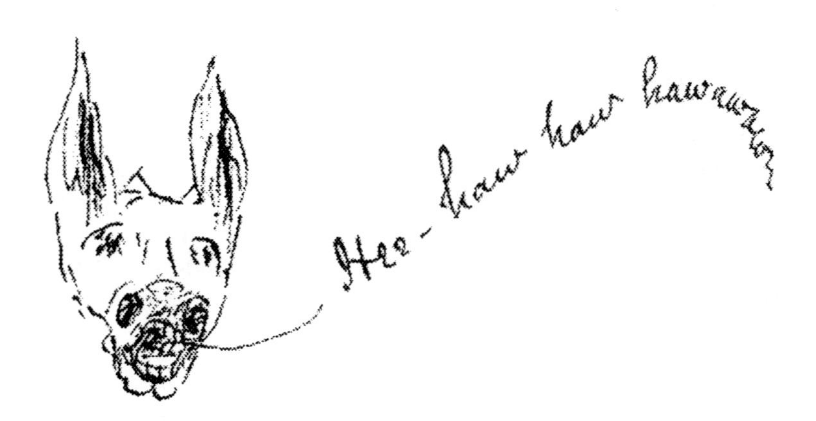

Ranyard once travelled on the continent with the Webbs, and there is no doubt that they saw him as an extended family member. 'Think of us at Xmas,' wrote Thomas, 'and come to pay us a visit, if only for a week ... for you seem to belong to us now, in some shape.'[21]

Thomas continued to lecture in Cheltenham throughout these years, sometimes staying in Cheltenham for four weeks, and going home only at the weekends to take services. His lectures at Cheltenham Ladies' College included lessons on optics (one of the subjects set for examination by London University[22]), but also included esoteric subjects and sometimes observing. In 1875 he commented to Ranyard: '70 to 80 girls looked at ☽ [the Moon] & ♄ [Saturn]. Very pretty & very sharp – (I mean the lights! Though not disparaging the Girls!).'

Both Thomas and his father gave talks to the Hereford Literary and Philosophical Society.[23] It is also probable that Thomas, always ready to popularise and educate, gave many talks and lectures in nearby Hay over the years, though there is only one lecture – on spectrum analysis – mentioned in a letter. Clearly, Webb was not expecting a lively audience, for he said that it would be 'more likely to be agape with sleepiness than curiosity'.

In 1859 Thomas had mentioned to Ranyard that his father, then eighty-three, had suffered a long illness. It was probably this which led to John's retirement from Tretire and his resignation from his benefice at Cardiff. Thomas and Henrietta invited him to come to live with them at Hardwicke. Remembering that Thomas had lived mostly in his father's house for forty years and worked for him as a curate, we can admire the respect and filial love when reading part of the letter that Thomas wrote:

> My dearest darling Father, We fully expected to have heard this morning, & sent into Hay for the letters. But none from you – we have no doubt that you are giving Mrs Penoyre's kindness due consideration – & most earnestly hope the result will be Yes. Just think how very happy it w'd make *them* & *us* to have you here! And I don't see why it sh'd cause *you* the least diminution of comfort except the mere moving of books, which I w'd look after. Pray do come.[24]

With the agreement and probably the financial support of the Penoyres, the house was enlarged and gentrified. Thomas Henry Wyatt,[25] Henrietta's cousin, added a fine decorative porch and full-height bays to back and front; but the major addition was a new wing into which the servants' quarters were moved and the domestic offices rearranged so that John could have a bedroom and study of his own. In September 1860 John duly moved in with his multitudinous papers, books and letters – but not without some anguish. Thomas, writing to the secretary of the Royal Astronomical Society at this time, commented:

> We are all in the miseries of packing and moving house and turning over the hoards of many, many years – often a source of painful recollection – and I shall be glad when it is all over, and my dear father, whose household is being broken up, is fairly settled with us at Hardwicke.[26]

The three adjusted and balanced their lives together. Indeed their relationship seems a model for those who share (admittedly commodious) surroundings with their family. John died at Hardwicke vicarage on 18 February 1869, in his ninety-third year and still very much in his right mind. The next day Thomas asked the vicar of Clifford to officiate at the funeral at Hardwicke church, and later the Webbs put up a plaque in the sanctuary recording his involvement with that church, and added lovingly: 'Senex quam rarus' (a rare old man), like a rare old wine.

There were other family members who enjoyed the Webbs' loving attention. Louisa, Oswald and Helen Wyatt were ten, nine and seven years old respectively when the Webbs moved to Hardwicke, so it was probably in the early 1860s that their parents, in the fashion of the time, gladly shared the care of their eldest children with their aunt and her husband. As they grew older it appears that the two girls may have spent more time in Hardwicke than in the family home at Troy House, and were almost informally adopted.

We cannot prove that the Webbs ever wished to have children of their own, but they were obviously more than ready to fill the valuable role of adult friends to nephews, nieces and young astronomers. In many ways during their life at Hardwicke the couple's concern and love for their relations, friends and servants created an extended family group. By showing affection, encouragement and concern for so many, one can liken them to the best of childless couples who add a dimension to many young lives, untrammelled by the demands of their own nuclear family.

A young curate also shared this attention. In 1865 the Revd Francis Kilvert was appointed curate at Clyro – a village across the river from Hay. His diary – probably kept from 1869 to his premature death ten years later – recorded a vivid and detailed picture of life in the locality.[27] He attended Hardwicke parish events and social occasions and met some of the regular stream of visitors. He recorded that Thomas Webb walked over to Clyro to hear him preach, and in return Kilvert visited sick parishioners in Hardwicke when the Webbs were absent. It could be a lonely life for a curate in a parish, and Kilvert must have highly valued the friendship extended to him by the Webbs and the support that the senior colleague gave him.[28]

One parish event described vividly by Kilvert was a bazaar in aid of

Hardwicke Vicarage in 1867. Note the telescope standing beside the figures of the Webbs at the front door.

Home Missions held in the grounds of Hardwicke vicarage on Monday, 29 August 1870. Two of the standard tomes of the time on the duties of the country parson were firmly against bazaars – one considering them to be a distasteful way of raising money, the other as a 'mere vanity fair'.[29] Neither Thomas nor his Bishop was of this opinion, and so Thomas made all the arrangements, despite some misgivings:

> The day for the Bazaar is *29 August*. Day is rather unfortunate as just preceding Hay Flower Show – so those people will not come – and in the middle of harvest – but it was fixed by the Bishop so there is no help for it but in getting good interesting speakers. About this I am sorely anxious.[30]

Kilvert arrived at Hardwicke to find:

> The Bazaar was in full swing, the tent very hot and crowded. Everybody was buying everything at once. The Hay Volunteer Band banged and blasted away. Persons ran about in all directions with large pictures and other articles, bags, rugs, cushions, smoking caps, asking everyone they met to join in raffling for them.[31]

The Bishop of Hereford 'spoke readily but not eloquently for an hour' – a lesson on how not to do it at a fund-raising event in any age! Kilvert continued:

> He is not a born orator and does not excite enthusiasm. But he was so engrossed in his subject that he forgot all about time and Mrs Webb had to give him a hint that the carriage was waiting to take him to Hay Station and that if he stayed a few more minutes he would lose the last train.

Despite the difficulties before and on the day, Thomas wrote triumphantly to Ranyard: 'The result was beyond all possible anticipation – about £60. The expenses cannot exceed £15.'[32]

Thomas had many dealings with the children of Hardwicke. Much of what he did was in the line of pastoral duty: school visits, encouraging the choir, arranging outings, and no doubt organising the Sunday school. However, there are many hints that they were duties enjoyed both by him and his wife. It may have been obligatory for a vicar to give the church choir an annual party, but perhaps not so usual to put up flags on poles and banners declaring 'Welcome to Hardwick'. Again Kilvert captured the spirit of the occasion on 14 July 1870,[33] and described how the children were sat at a table 'spread with a white table-cloth and adorned with vases of bright flowers' on the back lawn.

After dinner Oswald Wyatt sent up hot air balloons, and the children chased them across the fields and then returned to receive their prizes. These were no mere tokens, and included a photograph frame and a microscope. Thomas clearly valued his choir of children, describing them in a letter as 'the élite of the parish'.[34]

Thomas would walk from the vicarage across the fields to the elementary school, which was founded in 1836. He was well liked by the children, and some, at least, acted as reporters of 'natural phenomena' which they had seen. Thomas proffered some teaching in religious knowledge and no doubt spoke of the wonders of the heavens to the young pupils. The school was commended by inspectors for the standards reached in religious knowledge as well as the 'three Rs' of reading, 'riting and 'rithmetic. As the agricultural depression of the 1870s deepened, Thomas was involved in a free soup scheme for the poorest children.[35] The school log from 1874 onwards has many references to the Revd Webb, and records when the children of the Hardwicke choir were absent to attend a choir festival at Hereford Cathedral or special services at Hardwicke, and above all to enjoy the dinner or treat at the vicarage. Thomas occasionally gained wry amusement from his visits to the school:

> Our school children (under gov't) are now to be taught Scientific Grammar. One of them being asked about the number of Genders, answered there were Three Ganders – I believe so too – & who was the Goose!! Not the poor child![36]

Thomas and Henrietta had not come to a wealthy parish, but through their birth and connections the houses of the 'gentry' of the surrounding area were opened to them. They were on visiting terms with the Penoyres at The Moor and the de Wintons at Maesllwch Castle, besides having titled relatives and friends to stay at the vicarage. They were also part of 'the usual set' of gentry and clerics who lived in and around Hay. Kilvert often described social occasions in detail, with mention of the participation of the Webbs:

> Tuesday, 12 July 1870 ... Walked to Clifford Priory ... A crowd in the drawing-room drinking claret cup iced and eating enormous strawberries ... [There] were the usual set that one meets and knows so well. Dews, Thomases, Webbs, Wyatts, Bridges, Oswalds, Trumpers, etc. Everyone about is so pleasant and friendly that we meet like brothers and sisters. Great fun on the lawn, 6 cross games of croquet and balls flying in all directions. High tea at 7.30 and croquet given up ... More than 40 people sat down. Plenty of iced claret cup, and unlimited fruit, very fine, especially the strawberries.[37]

Inevitably there was occasional friction. In 1874 both Frances Penoyre and her daughter had died, and the new residents of The Moor were, according to Thomas gossiping to Ranyard, generally disliked by the neighbourhood, the wife being 'unpresentable'.[38] Henrietta would not call upon her because 'the lady [was] so intensely vulgar', though Thomas himself occasionally called on business, and there was disquiet over the felling of trees near to the vicarage.[39]

In the small community there was plenty of social small talk, rumour and gossip. It is possible that Henrietta was content with such a quality of social intercourse and frisson. Is there the sound of lively chatter when Thomas was asking a legal question in a letter to Ranyard about parties in the parish and ownership of a well?: 'I have stated the case – not as precisely and technically as I ought – for there have been three ladies in the room.'[40]

Thomas may have appreciated more frequent opportunities for intellectual exchanges. In 1876 he wrote wistfully to Ranyard, urging the latter to come and stay: 'No one to speak to here except about timber, game, balls, paupers etc. You will quite refresh us.' It is most likely that he occasionally met the Revd Henry Cooper Key[41] – another clergyman astronomer, who lived at Stretton Sugwas near Hereford – for when Key died Webb wrote that he missed him very much. Undoubtedly he went to see George With, the Master of the Bluecoat Schools in Hereford, and enjoyed conversation about specula. Although he never mentioned his own share in the progress of With's work, the latter was thankful for his encouragement:

> What a debt of gratitude I owe to our friend. It was he who encouraged me in my observational astronomy. It was he who gave me my first hints on speculum working, for he had worked small metallic specula for himself; and it was he who first brought my poor work into notice.[42]

Nonetheless, lack of time, slowness of travel, and the absence of the forms of communication now available, makes it difficult for us to realise the isolation which Thomas and many others must have felt. He assuaged it by his voluminous postbag, but this could not be as immediate and stimulating as face-to-face conversation.

Other young astronomers provided opportunity for such conversation in the later years of Webb's life. One visitor was Herbert Sadler,[43] whom Webb thought would be 'one of the first observers of the rising generation'.[44] Sadler was of some assistance to him when Webb was preparing a new edition of *Celestial Objects*. Another was Thomas Henry Espinall Compton Espin,[45] who was a pupil at Haileybury College, in Hertfordshire, when he contacted Webb in 1876. One of the masters had a small observatory and had introduced the boy to astronomy. He proved to be an apt pupil, and in May 1876 he had a letter published in the *English Mechanic* reporting his observations of θ Orionis, this being the first of a multitude of letters, articles and papers which appeared over his name throughout his life. In November 1879 he visited the Webbs at Hardwicke.[46] By then he had already been elected a Fellow of the Royal Astronomical Society and had gained entry to Oxford. Perhaps one topic of conversation was the desirability of forming an association of amateur astronomers. Certainly, Espin wrote a letter to the *English Mechanic*, published early in 1880:

> I have letters which show nearly the same feeling of isolation that I used to experience when first I commenced star-gazing, and it would be interesting to know whether some society could be formed for drawing closer ties between us brother star-gazers.

A year later the Liverpool Astronomical Society was formed, with Espin as a founder member.

In early 1883 Webb wrote to Ranyard: 'We have had T.E. Espin here for the inside of a week. He is a very charming fellow and I think has a bright astronomical future.'[47] Webb was not only a good judge of character and ability, but was prepared to back his hunch by appointing Espin (not yet twenty-five years old) as an executor to his revised will. Espin no doubt asked whether Webb would accept the honour of becoming a Vice President of Liverpool Astronomical Society. Webb agreed, and was elected on 23 February 1883.

Henrietta, unencumbered by either a career or many household chores, developed skills in more artistic directions. There is evidence that Thomas liked to sketch when time allowed, but it was Henrietta who developed her talents in painting. In the pages of Kilvert's diary and Thomas's letters there are many mentions of her work, often executed during holidays abroad. Henrietta would offer her paintings to raise money for charity. Thomas speaks of the picture of the 'Blumlis Alp from Lake Thun' and of a 'lovely table screen of my wife's painting',[48] and Kilvert wrote that at the Hardwicke bazaar:

> There was a series of water colour sketches of Alpine scenery, Monte Rosa, the Matterhorn, etc. beautifully executed by Mrs. Webb, 10/- each. I got up a raffle of ten people 1/- each for a beautiful drawing of the Schreckhorn one of the series. Mrs. Powell of Dorstone drew the prize.[49]

Sadly, despite some investigation, no paintings by Henrietta have been found except for the painted door of the drawing room in the vicarage. Sprays of leaves and blackberries, painted in oils, fall artistically across

Drawing room door, painted by Henrietta Webb, at Hardwicke Vicarage.

the top panels, and in the smaller panels below are vignettes of ferns and a corncrake standing amongst reeds.[50]

It became fashionable amongst the ladies of Hay society in the 1870s to decorate their rooms with their own painted designs. Mrs Crichton of Wye Cliff, Hay, decorated the walls of her boudoir. Did she decide to emulate and surpass Mrs Webb, or was it the latter who was inspired to copy her neighbour? It is probable that Henrietta painted, in watercolours, a postcard of Hardwicke church, after which Thomas sent photographic copies of the painting to his friends.

Henrietta became particularly absorbed in photography. In the 1850s the introduction of the *carte de visite* – a visiting-card sized photograph – became a craze, and Henrietta was one of many to take up the hobby. Thomas shared her interest – at least theoretically – for he made copious notes on photographic processes and recipes for varnishes for magic lantern slides.[51] Ranyard was asked to acquire clear varnish for protecting negatives and photograph frames. A year later Webb wrote:

> My wife sends you even better photographs tho' not yet satisfied with her papers for printing – a specimen of a photographed fern ... She wishes to have your opinion as to whether such things (i.e. a little book containing 2 or 3 dozen of them) would be likely to sell for charity.[52]

Thomas and Henrietta took great interest in their garden, which had 'delightful lawns and rockeries'. Clearly unworried by any environmental concerns, they had a fine collection of Alpine plants which they had gathered in Switzerland. Thomas urged Ranyard to come to see 'how marvellously the Swiss collection flourishes and blooms here'.[53] These plants were still flourishing in 1912.[54] The Hay Gardening Society show awarded prizes to Mrs Webb for a garden with a 'part-time gardener'. A large conservatory attached to the front of the house and a heated greenhouse in the garden contained, according to the sale catalogue, more than three hundred plants at the time of Webb's death. Also in that list were eight hives of bees and a new patent galvanised honey extractor. Although they had James Willis and their part-time gardener to do most of the outside work, one suspects that Henrietta was also actively occupied in the garden.

Holidays and other times away from the parish required careful consideration of how the clerical duties would be covered. If Thomas was away for just a few days then he would ensure he would be back in time to take services at Hardwicke on Sunday, by returning on a Friday and departing again on Monday morning. Alternatively he would appoint another clergyman to stand in for him. He once did this for six weeks in succession when staying in London.[55]

Visits in Britain were made to Devon and various parts of Wales – Tenby, in Pembrokeshire, being a favourite destination. The main annual holiday was spent on the Continent – often in Switzerland, Germany or France, and sometimes Italy – and lasted for around six weeks. More than once they took their nieces with them. Since the party also included Henrietta's maid, it was quite an entourage that travelled by carriage to Gloucester and by train to Paddington, London. There they would stay for a few days, and then catch the train at Charing Cross to Dover. A voyage on the Channel packet to Calais or Boulogne followed, and then trains and carriages conveyed them on the Continent. Sometimes they would move on 'by a sudden resolve' to another city or country, and would visit foreign astronomers. In 1872 Thomas wrote to Ranyard from Albrück:

> I am starting this letter at a Gasthaus near a small station on the Basel to Schauffhausen line ... where the filthy, smoky engine of the Grand Duke of Baden is to pick us up in two hours' time to deposit us this evening

> within sound of the giant fall ... [We had] a very smooth passage – a comfortable house at Brussels – a dull journey to Trives (but there most interesting Roman remains) a nasty filthy rail not far from the scenes of bloodshed at Metz and Forbach, to Mannheim – then a pleasant Sunday at Freiburg – and since a charming carriage journey through part of the Black Forest by the great Byzantine church in the strange seclusion of St Blasien to this place.

They admired the Chûte de Rhin, and went on to Lucerne. After that their plans were

> ... a little bouleversé – and the post is coming – and we are writing in a hurry – and it is charmingly quiet – and we enjoy it greatly – and we write in kind love to you and yours – and I remain, My dearest Arthur your affectionate old friend.[56]

Interested in both people and places, they chatted to local inhabitants and to hoteliers and guides. In Ilfracombe, Devon, they spoke at length to 'Ann Berry of Hale, a very intelligent and well disposed donkey driver' about smuggling, wrecks, and weird lights. A guide whom they had hired in Switzerland felt sufficiently friendly with Thomas to write for advice when he thought of sending his daughter to England to improve her English. Thomas could, however, be a little scathing about some foreigners:

> I really have a great respect for genteel Germans. But vulgarity – the same in its essence everywhere is especially unattractive when rolling about in a cloud of smoke ... [referring to a pension] the house is changed, & the management more pretentious. I think Switzerland is over doing the thing, & being overdone – they will be done up in the end.[57]

Upon their return the Webbs shared their experiences with their neighbours and parishioners. Kilvert mentions a lecture on Switzerland given by Mrs Webb at the school. He averred that it was: 'Most interesting, especially about St Bernard's Hospice, the dog coming down the Pass to meet the travellers and the tolling of the bell to announce their coming.' Such accounts to those who have been unable to enjoy holidays themselves are not always welcome, but perhaps Henrietta was a lively speaker. Her slides were 'encored' upon one occasion, 'and not much wonder', added her fond husband, 'for they were charming'.[58]

As the Webbs grew older, physical ailments began to feature in the correspondence. Henrietta went to Cheltenham for a course of

Turkish baths, and in 1870 Thomas took her to Bath for 'warmth and regular treatment'. When describing earth tremors in his later notebooks he mentioned Henrietta's rheumatism. It might also be wondered whether she was pleased to be woken up to corroborate his observations.[59]

In 1874, while in Paris, Thomas was seriously ill with a 'stoppage of the bladder'. After difficulty in finding a specialist and a night when Henrietta despaired of his life, an operation saved him. He was unwell for some time, and was most thankful for the efficiency and tenderness of his wife's nursing. While recounting this experience to Ranyard he moralised upon the 'vanity and emptiness of all more intellectual pursuits' and the 'unutterable folly' of deferring repentance, adding that God was his saviour and his refuge. Perhaps there was indirect advice here for Ranyard, who was less certain of his faith. In 1880 Thomas also had a distressing condition of acute eczema on his feet. It continued for a year, but he made light of it to Ranyard, illustrating his letter with small sketches.[60]

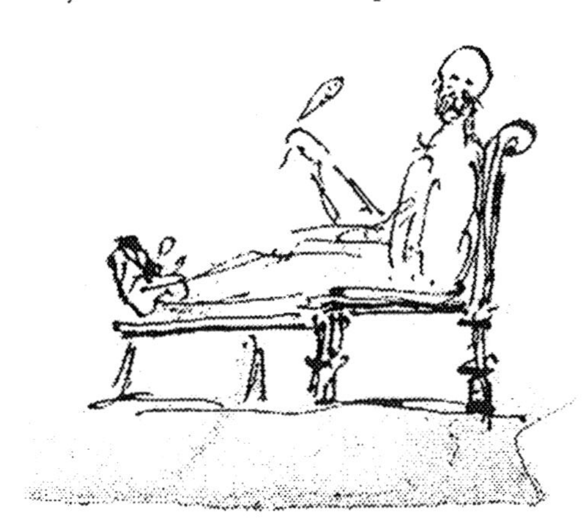

In 1880 – when Thomas was seventy-four and Henrietta was sixty-four – a growing sense of mortality began to creep into his correspondence: 'The Mistress sends her kind love but we despair of your ever coming to see us before we die'; and 'Very many thanks from an old friend with a sharp scythe [Father Time] poking into his back.' He asked Ranyard for help with his book on optics, saying that he knew the answers, but what with the 'pappy condition' of his brain and surgical and other bothers he could not remember, and disparagingly signed himself 'Turnip Top'. The couple nevertheless continued to take long holidays, visited Cheltenham to give lectures, and conscientiously pursued the daily round.

Despite advancing age Thomas was progressive in matters of women and money. Towards the end of 1882 Henrietta received a sum of money from the will of her sister Emma. Spurred on, no doubt, by the Married Women's Property Act (passed that year), he wanted to ensure that his wife could do as she liked with her own money – something denied to married women up to that time. He wrote to Ranyard: 'Is there any way which gives her sole right to this, so that I have no control over it . . . and she can do as she pleases with it entirely – which I wish her to.'

On 7 September 1884 Henrietta died suddenly from a stroke. Thomas wrote the following day to Ranyard:

> My dearest Arthur, None will more faithfully sympathize with me than yourself in this grievous affliction. My dearest wife has left me to join your good mother and the Saints in Paradise. This took place from apoplexy yesterday evening. Pray for me – you will be heard for Christ's sake. Your very loving friend . . .

Henrietta was buried at Mitchel Troy, near Monmouth, where, wrote Webb,

> . . . we were married 41 years ago (blessed be God for that!). She had chosen that ground 7 years ago & was borne to it *covered* with her earthly treasures of flowers, disposed in wreaths & crosses – no pall.

The memorial plaque in Hardwicke church speaks of her as 'A woman of very rare gifts and of a most generous heart.'

Thomas's life was muted by sorrow until his own death eight months later. He thoughtfully gave away items belonging to Henrietta in remembrance of her. Ranyard had a 'good binocular she travelled with, a new railway reading lamp', and godchildren were given mementos – a ring, a crystal picture and a musical box. He continued to observe – commenting to Ranyard that he had not 'forsworn astronomy'[61] – and conscientiously cared for his parishioners and servants. He also sorted his papers, and made 'quiet and gradual preparation for the general sale which must some day or other empty the poor old vicarage'.

The love which he and Henrietta had shown to their young friends was reciprocated at this time, and one after another came to be with him. His relations too supported him, and he visited the Wyatts and had cousins to stay at Hardwicke.

On 12 March 1885 he altered his will with the Revd T.H.E.C. Espin, Revd Charles Samuel Palmer (vicar of Eardisley) and George With as executors. On 6 April he attended what was to be his last Vestry meeting during Easter week, and took the festival services. A month later, Fanny Dew – sister of the Dews of Whitney Court on the north bank of the Wye, and a long time friend – came to be with him. He took to his bed on 18 May, and died the following day.

Two letters written a few days after Thomas Webb's death are demonstrative of the esteem and love of local people, and buttress the affectionate and admiring obituaries published in newspapers and official journals. The first is part of a letter from Louisa Bevan, wife of the vicar of Hay, to her brother the Revd Richard Lister Venables, the vicar of Clyro. She wished him a happy birthday, and continued:

> Fanny Dew came to Hardwick Monday week having heard Mr Webb was ill; she could not persuade him to see a doctor till Friday and the Tuesday morning he died. No one in the house but Fanny and the servants, but the Wyatts came Tuesday night. He suffered much of late but was so unselfish and patient that he never complained, his wife's loss to him was quite irreparable, he depended so much on her for everything. We have lost a very kind friend and neighbour.[62]

Fanny Dew, in a letter, commented:

> Helen Wyatt & I took a turn in this lovely little garden all bright with gentians & spring flowers – so sad to think that the dear inmates of Hardwick are passed away. They were such loving kind friends to me ever since they came here. We can hardly believe that every thing in this pretty house & garden must be swept away by a sale shortly. I feel we shall not see his like again, so cultivated a mind, & talented in many ways & yet such a meek Xtian spirit.[63]

Thomas was buried beside Henrietta, at Mitchel Troy, on 23 May. At Hardwicke a plaque was placed in the chancel beside that which Thomas had put up in memory of his beloved wife. It reads:

To the much loved memory of
Thomas William Webb, M.A., F.R.A.S.,
Prebendary of Hereford and Vicar of this parish
A man holy and humble of heart
Who followed the Master in all holiness and meekness
and now rests in Him after his labours.
He fell asleep May 19th, 1885
Aged 79 years
And is buried at Mitchel Troy near Monmouth.

Thomas's will had been drawn up with care and forethought.[64] He left money to servants, young friends, relatives, executors and his astronomical legatees. Property inherited from his father, and the proceeds of the sale, was dutifully left to a cousin. The rest of his estate – some £12,000 – was left to charities connected to the Church. There were no surprises. Other godchildren and children of local friends benefited from the Webbs' benevolence. Espin was given a legacy of £100 in acknowledgement of his work as executor which, incidentally, he put towards the cost of a 17¼-inch reflector by Calver.

In the custom of the Church of England the vicarage had to be vacated quickly for the new incumbent. The executors – having no doubt taken the astronomical notebooks and papers which Webb had given to them – arranged for a sale of the contents, which was held at the vicarage on 20 and 21 July 1885. Three days later, 3,000 books and a quantity of silver and silver plate were auctioned in Hereford. The logbook at Clifford school reports that the children were very much interested in the sale, and many stayed away from school in order to attend it. It is known that several parishioners bought items both for use and as mementoes. Since the auctioneer sold the furniture and other household goods in the room where they had always been located, it is not difficult, with the catalogue in hand, to visualise the various rooms in the house and sense the ambience of the place.[65] Suddenly, lot by lot, that ambience was dispersed. The Victoria Sociable carriage with two dark brown horses, a park Phaeton with a 'handsome bay horse having fine action', a Newtonian reflecting telescope and an observatory – 'to be sold as one lot' – furniture, pictures, clocks, pianoforte, chamber organ – all were bought and taken away. Hardwicke vicarage, which had seen almost thirty years of productive work and happy marriage, became an empty shell.

HARDWICK VICARAGE,

HEREFORDSHIRE,

ABOUT 2 MILES FROM THE TOWN OF HAY.

CATALOGUE

OF THE

Horses, Carriages, and Appointments; Live Stock, Agricultural and Garden Implements, Greenhouse Plants;

Newtonian Telescope and Observatory;

THE WHOLE OF THE

HOUSEHOLD FURNITURE,

PICTURES, GLASS, CHINA, LINEN, & OTHER EFFECTS;

Valuable Chamber Organ, Pianoforte, &c.

TO BE SOLD BY AUCTION, BY

MR. O. SHELLARD,

On the Premises, July 20th, 21st, and 22nd, 1885;

AND ALSO OF ABOUT

500 OZS. OF SILVER,

A QUANTITY OF SUPERIOR PLATED GOODS,

AND MORE THAN

3000 VOLS. OF BOOKS,

WHICH WILL BE SOLD AT THE

Hereford Corn Exchange,

ON FRIDAY, JULY 24th, 1885,

N.B.—The sales at Hardwick Vicarage will commence at 11 for 12 o'clock punctually, and at the Hereford Corn Exchange at 12 for 1 o'clock punctually.

The cover of the sale catalogue of Hardwick Vicarage and its contents, 1885.

Notes and references

1 Webb to Ranyard, August 1872.
2 T.W. Webb, letter to the Secretary of the Royal Astronomical Society, 5 January 1857, RAS Letters.
3 In Webb's time, Hardwick was spelled without an 'e', which was appended to the name at the beginning of the twentieth century, although both versions are still used.
4 See Chapter 4.
5 A. Mee, *Observational Astronomy: a Practical Book for Amateurs*, second edition (Western Mail, Cardiff and London, 1897).
6 T.W. Webb, 'Observations of Meteoric and other Natural Phaenomena, 1867–74', MS in Hereford City Library.
7 See Chapters 7 and 8.
8 G.F. Chambers, obituary of T.W. Webb, *Nature*, **32** (1885).
9 See Chapter 7.
10 T.W. Webb, quoted by Arthur Mee in *Observational Astronomy* (n. 5).
11 See Chapter 15.
12 Ibid.; see also Bibliography.
13 Mee (n. 5).
14 Arthur Cowper Ranyard (1845–1894): barrister; Fellow of the Royal Astronomical Society, 1863; Secretary of the RAS, 1874–80. In 1870 he was Assistant Secretary of the organising committee for the expedition to Spain to observe the total solar eclipse. In 1879 he compiled and edited a massive RAS *Memoir* recording all observed eclipses up to 1878.
15 Webb to Ranyard, 1 June 1858.
16 A reference to George With (1827–1904) – one of the first exponents of silver-on-glass mirrors.
17 See Chapter 8.
18 Webb to Ranyard, 12 June 1859.
19 Webb to Ranyard, 25 November 1863; and see Chapter 8.
20 Webb to Ranyard, 14 August 1879.
21 Webb to Ranyard, 24 August 1865.
22 See Chapter 15.
23 The Hereford Literary and Philosophical Society was founded in 1836. It had various names and fluctuating fortunes, gradually faded away, and closed in 1869. Both Thomas and John Webb were members, and at one meeting some of With's telescopes were displayed.
24 MS J12/IV, Hereford Record Office.
25 Thomas Henry Wyatt (1807–1880): architect, Hackney District Surveyor, President of the Royal Institute of British Architects, 1870–73, Gold Medalist, 1873. Designed, amongst other buildings, Knightsbridge Barracks, the Adelphi Theatre and Usk prison, and restored many churches.
26 T.W. Webb, letter to the Royal Astronomical Society, 13 September 1860.
27 W. Plomer (ed.), *Kilvert's Diary: Selections from the Diary of the Rev. Francis. J. Kilvert, 1870–79* (Jonathan Cape, London, 1938–40.)

28 Within living memory, an old lady – having been asked what her family had thought about Kilvert – said disdainfully: 'Think about him? We did not think about him. He was only the curate.' (Told us by David Lockwood.)
29 O. Chadwick, *The Victorian Church*, 2 vols. (Oxford University Press, 1966–70).
30 Webb to Ranyard, 27 July 1870.
31 Kilvert, *Diary*, 29 August 1870.
32 Webb to Ranyard, 10 October 1870.
33 D. Ifans (ed.), *F. J. Kilvert, Diary, June–July 1870* (National Library of Wales, Aberystwyth, 1989).
34 Webb to Ranyard, 12 October 1871.
35 J.A. Millardship and J.K.A. Morris, *History Of Clifford School* (privately published, 1986).
36 Webb to Ranyard, 20 January 1876.
37 Kilvert, *Diary*, 12 July 1870.
38 Was this merely because the lady had been an actress?
39 Webb to Ranyard, 29 June 1870.
40 Webb to Ranyard, 3 March 1876.
41 See Chapters 7 and 8.
42 Letter from G.H. With to Arthur Mee, quoted in the latter's Memoir of T.W. Webb.
43 Herbert Sadler (1856–1898): Fellow of the Royal Astronomical Society, 1876. Industrious amateur astronomer. Frequent contributor to journals. Formed, with E. Neison, the Selenographical Society in 1879.
44 Webb to Ranyard, 24 March 1875.
45 See Chapter 7; see also A. Brown, *The Life and Work of T. H. E. C. Espin* (unpublished thesis for MSc, Durham University, 1974).
46 Webb to Ranyard, 11 and 19 November 1879.
47 Webb to Ranyard, 2 February 1883.
48 Webb to Ranyard, 12 July 1872.
49 Kilvert, *Diary*, 29 August 1870.
50 The authenticity of this painting was established in a paper in *Journal of the British Astronomical Association*, **46** (March 1936), 188–9, in which Alfred Noel Neate describes a visit to Hardwicke in 1912: 'In the drawing room Mrs Webb's oil paintings continued to adorn the panels of the door, as indeed they do at the present day.'
51 T.W. Webb, 'Notebook 1', MS in Hereford Cathedral Library.
52 Webb to Ranyard, 8 August 1865.
53 Webb to Ranyard, 29 June 1870.
54 See n. 50.
55 Webb to Ranyard, 5 April 1865.
56 Webb to Ranyard, 11 June 1872.
57 Webb to Ranyard, 21 June 1869, from Züg.
58 Webb to Ranyard, 24 January 1872.
59 See Chapter 6.
60 Webb to Ranyard, 6 January 1880 and 16 March 1880.

61 Webb to Ranyard, 29 September 1884.
62 Llysdinam collection, MS in the National Library of Wales.
63 Ibid., undated, but in March 1885.
64 MS J/12/II/3, Hereford Record Office.
65 A copy of the sale catalogue is in the Pilley Collection in Hereford City Library.

Chapter 4

Man of the cloth

Mark Robinson

Working for work's sake, and that from the highest motives.[1]

At the beginning of the nineteenth century the established church of which Thomas was a member was in 'fat slumbers', privileged and unreformed. Many high-ranking clergy were enormously wealthy. Thomas once exclaimed: 'Why! the Bishop of Exeter has £121,000 per annum!'[2] – a splendid exaggeration to make his point. They were often absent from their benefices, whereas many curates were seriously underpaid. Tithes[3] – a form of tax that was strongly resented – were still often paid in kind.

As a consequence, Methodism and other dissenting sects flourished and fulfilled the needs of the ordinary people. Gradually, reforms were made. In 1836, tithes were commuted to a rent-charge on land, beneficed clergy were expected to live in their parishes, and clergy incomes were made much more equitable. Religious tenets throughout the century were questioned and discussed, and in the 1860s the basis of belief was further challenged by the theories of Darwin. These challenges caused many clergy to change their ways.[4]

The career opportunities open to young men of Webb's background were few by comparison to those a century later, and it could be inferred that Thomas had no great sense of calling to the priestly role and that he was doing no more than was expected of him. There may be an element of truth in this. Many sons of clergy followed in their fathers' footsteps at bidding, and it was a profession that could be adopted by a gentleman. On the other hand Thomas showed no inclination to follow other members of the Webb family into law or accountancy. In the case of the former he may have taken note of the many worrying times his father had when forced to seek legal opinion, usually on behalf of others. Nor did Thomas appear to express a desire to use his love of astronomy as a career. He lived at a time when the educated man could still combine study of a scientific subject with another calling which allowed some time and income for his hobby. All these and many other thoughts no doubt occurred to the young man in the years before he presented himself to the Bishop at Hereford.

He recorded one period of doubt about his calling to the ministry during his time at Oxford. At home in November 1829 he noted: 'Sunday 8th: Much oppressed by scruples eveng. & felt myself a condemned criminal – wretch that I am.' Ten days later his father came back from Gloucester, and Thomas wrote, no doubt thankfully: 'F. most kindly allowed me to defer resolving on a profession without making any enquiries.'[5] A wise father!

At Oxford, Thomas attended the lectures of Edward Burton, Regius Professor of Divinity and Canon of Christ Church Cathedral. In February 1830 he found them 'most excellent and impressive'. After the second lecture he spoke to Burton, and on 12 February wrote in his diary:

> Burton's 2nd excellent lecture in the big hall. Went and spoke to him afterwards about the opinion of the Ch. of E. as to Sententia Lata – he firmly discouraged such subtleties and he complained the question was too abstract, but when I came to the point of past life, he declared the Ch. of E. rejected no penitent from the Ministry – offered any further information, particularly civil, made me very happy.

Next day he wrote to Burton, and two days later attended

> ... Burton's 3rd glorious lecture on the Origin of evil – after wh. I went to his house by his desire & he gave me his ever memorable answer to my enquiries – praised be God. said he had read my letter w. much interest.[6]

Thomas cogitated for a few days and on the 17th 'remained long in prayer for direction. Went to bed late'. Next day he wrote a 'letter to my father containing my final decision, which I sent at night. Enter not into judgment w. thy servant O Lord!'. John Webb went to Oxford on 27 February, and all was 'settled providentially about Pencoed'. John had not applied pressure on his son to take holy orders, but at the appropriate moment had quietly made arrangements so that his son had the curacy at the neighbouring parish of Pencoed as his first appointment.

The diary entries for the days around the time of Thomas's ordination are among the longer and more emotional examples of his personal writing. Thomas had just bade farewell to Helen Duff upon her unexpected return to Scotland: 'I said amongst other trials I had much difficulty to bring my mind to this point i.e. my ordination.'[7] Thomas left Gloucester and went to Hereford for his ordination. In love, and in a state of uncertainty, it is perhaps not surprising that his record of events in Hereford are as much about mislaid luggage and missing robes as about the manner and solemnity of his ordination. Nonetheless he was impressed by the ceremony:

> 1 August ... Unequal Sermon by Pearson. had much fear and trembling & mixed feeling – yet as time drew on was, I thank God, supported & enabled to go thro' this solemn office w. comfort & pleasure. Thus was I ordained a Deacon. In defiance as it were of the powers of darkness – Glory be to thee O Lord most High – A day never to be forgotten – Grant, O Lord may it ever be remembered for good.[8]

A year later he was priested. Formal training or preparation for ordination was notable by its apparent absence. Other than the concentration on divinity at university, ordinands had only to pass simple tests of competence, and Thomas went to Hereford the night before the service in order to be examined the following morning:

> Dressed myself & went to the Palace. read Heber's excellent sermon on the duties of a Ministry – & was examined by Mr. Cook who stopped me in one place in my Latin where I was in too great a hurry, but was much pleased w. my Greek & burial service. he said it was an excellent examination. Then wandered about till 4h. & subscribed the Articles [The Thirty-nine articles of Religion], & got dinner at 6, having breakfasted at 7. Walked up Aylestone Hill, where a crazy boy had just run away w. a horse & gig. With very deep self-abasement & the most concentrated intensity of feeling, I was ordained Priest. Thanks be to God.[9]

It is doubtful that financial considerations weighed with Thomas at all in his choice of career, but he had the benefit of parental support. He could not have lived in this manner if he had had only his stipend. Curates often received less than £70 per annum, and as there was a surfeit of young men entering the Church, many curates never had parishes of their own. The stipend of beneficed clergy varied considerably from parish to parish. County directories of the period when Thomas became vicar in 1856 state that the living at Hardwicke was worth some £230 plus the rent-free house, although the produce or rent of the glebe land[10] would have contributed to his income.

Thomas William Webb – a *carte de visite* by a Leamington photographer.

Like his father, Thomas took paying pupils but there are only two references to them. In 1834 young Mr Symonds from Pengethley attended regularly to study the Greek testament and Sophocles. He came for the last time on 31 December 1835: ' … a nice good natured fellow but I am very happy to have done with him – he has taken up a great deal of time, obliged me to sit up a great deal & deprived my mother of much exercise.'[11]

In the 1860s Thomas also looked forward to supplementing his income from tuition. In a letter to his father he wrote that he had offered to teach the sons of the Clifford vicar:

> We get on well, thank God. In consequence of Mr. Fowler (of Glo'ster's) opinion that 5s would be fair for the Trumpers – which however, I felt rather too high for me to act upon, I was emboldened to ask 3s 6d per lesson, which did not seem to annoy them at all – and this, if it please God to give all parties health & free course & pretty fair weather to & fro, I hope to get about £35 or £40 annually, for which I hope I am very thankful.[12]

This commitment – the equivalent of four hours a week throughout the year – indicates Thomas's concern to enhance his income.

It cannot be said that Thomas and Henrietta had substantial wealth by the standards of the day; nonetheless, as Thomas was an only child his father's resources were available to him by way of money and gifts. John Webb had enjoyed various pecuniary advantages from both his family and his career. His wife had brought money to the marriage, he had been left money and property by his uncle, his mother also bequeathed property to him, and John himself seems to have invested wisely in property and shares. His salary from his appointment of Gloucester and his stipend from the living at Cardiff augmented the income of Tretire. The household which John and Sarah kept at Tretire was not lavish – John did not always keep a carriage – but it was comfortable. Neither father nor son seemed to lack ready cash when the needs of their churches were considered. Thomas not only provided an organ for St Weonards when he was a curate there, but also for Ganarew,[13] and John refurbished both church and rectory at Tretire at his own expense. Nor did Thomas and Henrietta seem to lack the means for the style of life which they enjoyed, though Thomas did not treat himself to the luxury of an observatory until after the death of his father. No doubt they lived modestly and within their means. Certainly Thomas, at his death, was able to leave substantial sums to the church.

Comments in several of his letters to Arthur Ranyard, however, add to the impression that Thomas was keen to augment his salary. Whether he regarded the money as an addition to his income, a necessity to pay for concomitant expenses, or as a matter of principle – the labourer being worthy of his hire – is difficult to tell. In 1874 he was undoubtedly pleased when the then patron of Hardwicke, Anna Maria Penoyre, left him a legacy of £500 plus £90 per annum to augment the living. Thomas said he thanked God for 'this most unexpected addition to our means'.[14]

He saw his long association with Cheltenham Ladies' College as another opportunity for the popularisation of astronomy and science, but he was paid at £1 10s a lecture plus railway fares, and undertook several series of lectures.[15]

In 1872, when the editor of *Nature* began to send him books to review, he commented: 'I don't much like the trouble of it but for two reasons it pleases me very much – it pays and it gives me a chance now and then of a quick hit at the adversary'.[16] And even in 1881, when Ranyard had put his name forward to lecture at Exeter Hall in London, he queried him about the length and quantity of lectures, adding 'I presume they pay?'.[17]

It is not easy to place the churchmanship of Thomas Webb in the spectrum of Anglican beliefs and doctrine, as he left few clues in his surviving writings. There are no sermons amongst them. He notes that early in his ministry the writing of sermons took considerable time, and he prepared them with care and thought. As for the sermons of others, his remarks rarely provide an insight of whether or not he found the subject matter thought-provoking. Whilst residing in Gloucester he frequently noted the music used at each service, but the sermons warranted the odd 'good' or 'magnificent'. On Easter Sunday 1830 he was filled with reverential fear: 'Spa church again at night. splendid but awful sermon, spoiled my night's rest. What vile corruption there must be in me.'[18] He had his moments of self-doubt – even to the extent of extreme emotional reaction when he mistreated a family cat:

> At night committed the blackest crime since my ordination, striking poor Ruff on the head with my foot – not intentionally however on the head – O Devil that I am! This spirit is the very spirit of manslaughter. Redemptor mundi miserere, miserere mei pectoris – Made me very uncomfortable – as well it might. She had provoked me to this by savagely fighting Mousey.[19]

His faith – which appears to have remained strong throughout his clerical career – was founded solidly on the Bible. Whilst in Hereford prior to his ordination as deacon he wrote:

> After dinner & coffee walked w Mr. Davis – a mulatto who had come w. me from Ross upon the coach. & was in a most distressed & agitated state of mind, longing for peace and reassurance. I could declare to him that I had passed thro' the same steps, & he was much comforted but amazed to find that I was disposed to set my face against high views & doctrine & against all reading of other religious books & Memoirs except the Bible.[20]

This attitude seems to have prevailed throughout his life. Nowhere in his later extant writings or letters is there any clear mention, for example, of the furore caused by Newman and the Oxford movement or of the controversy engendered by Darwin's *Origin of Species* which was published in the same year as *Celestial Objects for Common Telescopes*

(1859). In one letter written to Ranyard in 1865 he makes a general remark. He says that he hopes Ranyard will be active as a Christian and a scientist. Warning that Christianity and science were not converging to the Glory of God, and not even parallel – the divergence being the greatest evil of the day – he continued: 'Popery as bad as it is, will I am sure, be in the end found a far inferior mischief.'[21] 'An inferior mischief' – but he did see 'the Church of Rome' as 'the most glaring instance of a super structure of utter rubbish upon the true foundation'.[22] He was troubled, when young, by the Catholic Emancipation Act,[23] but his deeply held antipathy lasted into old age, as revealed by a letter to Ranyard in 1881:

> We are distressed very much by the perversion [*sic*] of our greatly loved niece at Chislehurst to the Church of Rome. We were especially attached to her – and she to us – and we had hoped to visit her this summer – but there is a wide and I fear irremediable separation now.[24]

He remained apparently untroubled by theological doubts and continued to rest his faith in a deep and simple creed, averring the need for 'earnest and humble prayer' and preferring to devote his intellectual energies to astronomy.

His observational books and other writings contain maxims, often in Latin, which illustrate his belief that his astronomical studies were an intrinsic part of his Christian belief and practice. Hearing that a Tulley telescope was to be given him by his father, he wrote: 'God grant that I may use it to his glory.' When it arrived he wrote in large bold letters: 'The good Lord give me the grace to make a Christian use of it – and that I may write upon it, Holiness to the Lord.'[25]

Espin refers to this in his 'Reminiscence' written for the fifth edition of *Celestial Objects*, and also quotes from one of Webb's observation books, the paragraph in Latin beginning 'Gratias D. O. M. refero'. In this heartfelt prayer Thomas firmly associates his astronomical endeavours with his belief in God the Creator, thanking God for the many benefits he has received through the years and for the pleasure he has enjoyed in looking at God's wonderful works. He says that, like Kepler, he is totally satisfied by beholding such glory and that he exults in all the work of God's hands.

He had uneasy feelings about other Christian denominations besides the Catholic church. Nonconformity warranted such entries as: 'Most pleasing news of the Ranters' meeting having been altered so as not to interfere w. church hours.'[26] If there had to be such a congregation in the area, then at least their services would not clash with those of the established Church of England! Worse was to follow: 'To Marie Powells, who to my surprize & Vexation had joined the Ranters –

Holy Trinity Church, Hardwicke, in 1867.

Holy Trinity Church, Hardwicke, probably painted by Mrs Webb, and reproduced as a photograph by 1870.

singular, that very evening after I had so much discussion w. her.'[27] Thomas was troubled, and 'thought much about Ranters. I was very low & dispirited at my unspeakable responsibility – & the terrors of eternity – O Lord deliver thy poor unworthy servant.' He was concerned that he had failed in his duty to Marie to keep her faithful to the true Church. Later Marie was seen again in the parish church. Thomas had found his lost sheep, but no exultation was expressed. Nearly a year later he commented on a meeting held with Ranter preachers,[28] but failed to note the purpose and results. In 1834 he was happy to go in 'the evening with M. to bury good old John Williams – grand Methodist funeral – 5 or 6 preachers'.[29]

At this time at least he found good theological reasons to avoid contact with the Evangelical wing of the Anglican church. This attitude is also reflected in the fair copy of a letter he sent to the editor of a religious periodical[30] on 3 March 1836. In this he queries the wisdom of those clergy who hold mid-week prayer meetings at which the Liturgy of the Church is not employed: 'This is what dissenters do [and] is it right to foster such a taste in our people who mistake the love of novelty for the love of God?' The next day he penned a somewhat briefer note seeking guidance from readers. The question related to baptisms in a large parish with a small and scattered population where few others would attend the service. He quoted the rubric in the Book of Common Prayer which required that the baptism should be in the presence of the congregation.

The pastoral ministry of Thomas Webb is well captured by Espin.[31] This vignette (fully quoted on p. 106) states that Webb would 'talk to his people of their everyday life, and local matters, making himself one of themselves, and imparting the sunshine of his life to theirs'. This was clearly a pattern of pastoral care which he had followed all his working life. In his first curacy at Pencoed, he recorded in his diary:

> The hardest day's work I have had for a long time. Morning pretty fair. turned out Pretty Polly & started without umbrella. During service rain came on; I had to go to Netherton; got half wet by the way in tremendous storms with my back against trees etc. Roads shocking. Baptized a child & sat some time; but it was no better – so I borrowed Davis's umbrella & set off home – a good deal of rain & dreadful wind. Went & fetched in my Pretty Polly – & then it was time to go again, which I did, with two umbrellas to Netherton, thro' some of the most dreadful weather I ever was in. Left J. Davis's umbrella with Christian. churched Kitty Williams, christened her child who roared from beginning to end, & received Powell's child – rain a good part of the way back. A very hard day – only sat down about 10m in 6½ hrs.[32]

Throughout his ministry – as curate to his father, and later in his own parishes – Thomas considered that visiting his parishioners in their homes – grand or humble – was an essential part of his work. His diary entries during the time that he was living at Tretire are full of references to such visits, whether on foot or on horseback. Should there be illness in a family, then he would visit frequently – no doubt 'laden with all kinds of little comforts'. His journeying could amount to seventy or so miles a week. In this way his ministry echoed that of many another country parson; for example, the Revd Francis Kilvert, in Clyro.[33] Kilvert refers to this activity, in his diary, as 'villaging'. Again it was conducted mainly on foot, and often covered the same distance as Thomas's travels.

Thomas's pastoral care also extended to legal questions. His letters to Ranyard, who was a barrister, are peppered with such queries. His gardener's complex problem over inheritance of property spawned two long, detailed letters of enquiry. After the passing of the Married Women's Property Act he was anxious that a woman parishioner, who had become engaged to 'a scamp of a butcher in Hay',[34] should not be fleeced of money that she possessed. He asked Ranyard if 'the mere existence of such an Act is a sufficient protection?'.

The parish priest was respected by his parishioners for the extent of his visiting and for his charitable giving, but the services on Sunday were his main obligation. Thomas recorded his first duty at Pencoed in 1830 thus:

> A day to be greatly remembered for I served Pencoed Church for the first time. Was not well, but I thank God got through the service easily – far more so than I had hoped. Was however sadly mechanical: partly owing to health. May this be to me O Lord, a new course of action & let me never forget I must pray for myself as well as the people.[35]

He regularly took the Sunday services at Pencoed as well as weddings, baptisms, churchings and funerals. Churching was the service at which the newly delivered mother gave thanks for her safe deliverance and preservation in the great danger of childbirth. He was upset that he was persuaded to church a woman in her own home, although a generous fee of 10s 6d was given, but he refused to conduct the baptism at the house. At a funeral he commented that there was a large gathering of '20 or 30 persons there' – probably men only, as it was often the custom for the women not to attend the funeral as they would be busy preparing the refreshments for the 'wake' that could follow the interment.

Services at church on Sunday followed a regular pattern comprising matins or morning service with a Sunday school for the children. Some ancient churches in the district have a porch entrance built large

A view of Hereford Cathedral from the north-east, 1831.

enough to hold the children's class, but at Hardwicke the patron funded a separate building for use as a Sunday school. It also included a coach house where their carriage and horses would be sheltered from the elements whilst they attended the service. In the afternoon the service of evensong would be conducted. Holy communion would not necessarily be a regular event, and was usually celebrated as part of the morning service. However, by the middle of the nineteenth century weekly communion was becoming a more regular feature.[36] All services would be taken in accordance with the rubrics of the Church and from the Book of Common Prayer – by then some two hundred years old.

Despite the fall in church attendances in the early part of the nineteenth century and the rise of Methodism and other nonconformist movements, the traditional idea of a village church as the hub of the community was still alive in many rural parishes in the 1850s. Hardwicke had the ingredients for this tradition: landlords of benevolence and longstanding in the presence of the Penoyre family, a vicar who assiduously carried out his duties, and a number of respectable farmers and their labourers to fill the pews. Webb's experience may have been similar to that of a preacher of 1852 at Lydiard in nearby Gloucestershire, who looked out of his pulpit and saw

> ... mostly snow-white smocks, here and there a blue smock with a pattern in white thread on the breast; each male with a red pocket handkerchief; two or three farmers in black coats, and women in dresses of various colours.[37]

Thomas was not entirely unreceptive to new ideas in patterns of worship. When he first became a curate the music in the majority of parish churches was performed by the parish band or not at all. Whilst there is record of his father giving musical instruments to the band, Thomas was keen to see an organ installed in Hardwicke church; but not for him a mere barrel organ[38] with a small repertoire of tunes. In February 1840 when he was at St Weonards, negotiations with his father and Walkers of London, Organ Makers, had reached a sufficiently advanced stage for Thomas to set out for London to place the order for a finger and barrel organ with extra barrels, costing £136 5s 10d.[39] At Exeter Hall – a principal concert hall of the day – Thomas heard 'a grand performance ... about 450 singers and the organ only ... no fiddles or trombones'. This was written in a letter to his parents; and we may forgive the sly reference to fiddles or trombones – a hint that perhaps the days of the church band were numbered. The organ was installed in August 1840, and on 8 September a festival service was held at which the Bishop of Hereford tried the organ himself. Anticlimax followed. The next Sunday: 'no organ service, choir not being ready'. No doubt the choir practised hard under the direction of Thomas during the ensuing week, for the following Sunday he recorded: 'First day of Choir Service. O how very delightful, unexpectedly so. Blessed be God for this very great and happy privilege.'[40]

In the running of his own church a parish priest of that time was usually his own master. Unless the patron of the living or local squires took particular interest in church affairs, there were few who could influence the decisions of the incumbent. Certainly there was no Parochial Church Council to placate, and there is no record of formality in the organisation of Hardwicke church affairs until 1882. In that year, immediately after Easter, Thomas convened his first Vestry Meeting, at which he formally appointed two Churchwardens for the ensuing year, in the presence of two householders, one of whom was the local innkeeper. Thomas faithfully followed this ceremony for the next three years, and it was not until the beginning of the twentieth century that a Parochial Church Council was formed. Thomas penned these early brief minutes of the meeting in his own firm clear handwriting that shows little sign of his advancing years.[41]

After more than fifty years as a priest, the Diocese of Hereford acknowledged Thomas's work by appointing him Prebendary of

Nonnington. In centuries past such appointments carried tangible benefits such as the income from farms and other properties owned by the Church. Though deprecatory about it, Thomas was obviously touched by the award of this 'long service and good conduct medal', as is evident in his letter to Ranyard:

> The Bishop has just presented me to a vacant Prebend in the Cathedral – no emolument and quite an inferior position to a Canonry – but still looked upon as a compliment – giving a position in the Chapter and a stall in the Choir. To me, lover as I am of music, and Church music above all, this renewed connections after so many years' interval in the Cathedral is very pleasant. Duty is only one sermon a year – and incidentally perhaps a little more presence at meetings of a Church character in Hereford. It was somewhat of a surprize – not wholly so – for I had been much more surprized (that *was* a surprize indeed – one of the very greatest I ever had) when the Bishop asked me to preach at his Ordination in Hereford at Christmas[42] – but my wife was quite taken aback at it – and to both of us it seems like a dream. I went yesterday to be 'collated' in due form – my 'institution' will take place at the Dean's convenience. In point of my long standing in the Diocese – I am so near the top of the tree, being now in my fifty-second year of service (without intermission except five years in Gloucester Cathedral) that perhaps if I never had such an offer it might have seemed intentioned – and however obscure I have been (especially so except for being asked to preach in the Cathedral like others in late years) I am not conscious of having given offence – so there is not much to be surprized at.[43]

The ministry of Thomas Webb may be measured against the ideals set out in J.J. Blunt's popular handbook, *The Duties of a Parish Priest*, published in 1856.[44] He should, said the author, read most mornings to gather material for the weekly sermon, which was the mainstay of his ministry. Thomas may not have scored too well against this criterion. After the difficulties of composing some sermons as a newly ordained priest, as recorded in his diary, there is little direct evidence that he spent undue time in preparation, even to the extent of admitting to using the same sermon at different churches on the same Sunday. Another criterion was that the priest should try to make his congregation take part in the service. Thomas ministered at a time when tradition was very strong that the congregation did no more than that which was decreed in the Book of Common Prayer. However, he used music to enrich the experience of the services in the church. Besides installing organs he formed parish choirs, encouraged congregational singing, and at Hardwicke he organised at least one choir festival.

Thomas admirably fulfilled the precept that the parish priest should know each parishioner well and build up the spiritual and temporal health of the community at large. Finally it was suggested that the parson should have strong influence with the local school and its young charges. Thomas's attention to his choir and the local school was exemplary.[45]

In summary, this man of the cloth emerges as a priest faithful to his beliefs and ideals throughout the fifty-three years of his ministry, largely uninfluenced by any controversy and reforming movements that engulfed some sections of the Church of England in his time.

Notes and references

1 T.H.E.C. Espin, 'T.W. Webb – A Reminiscence' in T.W. Webb, *Celestial Objects for Common Telescopes,* (fifth edition, 1893).
2 Diary, 12 June 1831 (see p. 21, n. 18).
3 Tithe: the tenth part of the annual produce of agriculture, being a payment for the support of the priesthood.
4 See Chapter 7.
5 Diary, 18 November 1829.
6 Diary, 15 February 1830.
7 Diary, 28 July 1830.
8 Diary, 1 August 1830.
9 Diary, 6 August 1831.
10 Glebe: land assigned to a clergyman as part of his benefice.
11 Diary, 31 December 1835.
12 T.W. Webb, letter to his father, undated (1863), Hereford Record Office.
13 It was a finger and barrel organ costing £125 (see n. 39).
14 Webb to Ranyard, 2 March 1874. The legacy was originally intended for John Webb. Thomas declared that it was an acknowledgement of what, in later years, had been painfully overlooked and ignored: his father's most valuable and faithful services to that estate.
15 I am indebted to Mrs K. Boothman, Archivist of Cheltenham Ladies' College, for this information.
16 Webb to Ranyard, 21 February 1872.
17 Webb to Ranyard, 26 November 1881.
18 Diary, 11 April 1830.
19 Diary, 18 January 1831.
20 Diary, 31 July 1830.
21 Diary, 31 July 1865.
22 Ibid.
23 See Chapter 1.
24 Webb to Ranyard, 13 January 1881.
25 Diary, 3 July 1834.
26 Diary, 8 September 1831. The Ranters were possibly Primitive Methodists – a nonconformist sect which originated around 1810.

27 Diary, 28 November 1831.
28 Diary, 14 August 1832.
29 Diary, 6 March 1834.
30 *British Magazine & Monthly Register of Religious & Ecclesiastical Information*, 36 vols. (London 1832–49).
31 Espin (n. 1).
32 Diary, 13 March 1831.
33 Kilvert, *Diary*.
34 Webb to Ranyard, 3 December 1883.
35 Diary, 29 August 1830.
36 O. Chadwick, *The Victorian Church*, 2 vols. (Oxford University Press, 1966–70).
37 Ibid. The last smocks disappeared in the 1860s.
38 Ibid. Chadwick cites Sir Frederick Gore Ouseley, who did much to raise the standards of Victorian church music. Ouseley said that such an instrument was 'praising God by machinery', and compared it to an oriental prayer wheel. The St Weonards organ had a barrel which played twelve tunes – no doubt for those Sundays when no competent organist could be found – but it was also built to a detailed specification for 'real' playing.
39 For all information on organs supplied to the Webbs I am indebted to J.W. Walker & Son, Organ Makers, Brandon, Suffolk.
40 Diary, 20 September 1840.
41 I am indebted to Hardwicke Parochial Church Council for the loan of the Vestry Book.
42 James Atlay, Bishop of Hereford (1868–95), held an ordination of curates and priests on 23 December 1881.
43 Webb to Ranyard, 28 January 1882.
44 J.J. Blunt, *The Duties of a Parish Priest* (1856; fourth edition, 1861).
45 See Chapter 3, pp. 44–5.

Chapter 5

The antiquarian and historian

Mark Robinson

My wife and I were both engaged in Civil War ('tho domestic peace) last night – I as usual verifying references – she examining loose papers.[1]

Being a dutiful son, Thomas Webb might have followed his father's bidding in choice of hobbies and interests. Whilst this is to a certain extent true, he soon tended towards the scientific rather than the literary and historic fields of scholarship and enquiry. Nonetheless, extant documents show that he was always ready to record a good historic yarn or take a journey to see a site of interest and to record that visit briefly but with his usual accuracy. Throughout his life he made notes on historic events or curiosities, and in the ten years after his father's death he was indeed dutiful in completing John Webb's unfinished works on aspects of the Civil War as it affected Herefordshire.

As a young man he was always ready to assist his father in his researches. He was as enthusiastic as the latter when, in 1825, John Webb recovered, from a local female herbalist in his parish, a stone bowl on which was carved a faint Latin inscription. Thomas deciphered 'Deo Triune Beccius Donavit', and the next evening worked on the last word 'ora' before resuming work on a brass tool for his specula. The inscription may be translated as 'To the God of the Three Ways given by Beccius' or 'Beccius dedicates this altar to the triune God'. The base was also recovered by the Webbs, and the complete Roman Altar (for that was what it was) was restored.[2]

In his youthful diaries Thomas noted historic remains which he encountered on his rides. In November 1826 he rode to Orcop 'and spied a grand tump – Festus dies'. He had not time to examine it at the time, but returned four days later and 'examined it all over with great pleasure ... 17 yards across the top: foundations of wall around it'. Three years later he took his father to the site for a more detailed inspection:

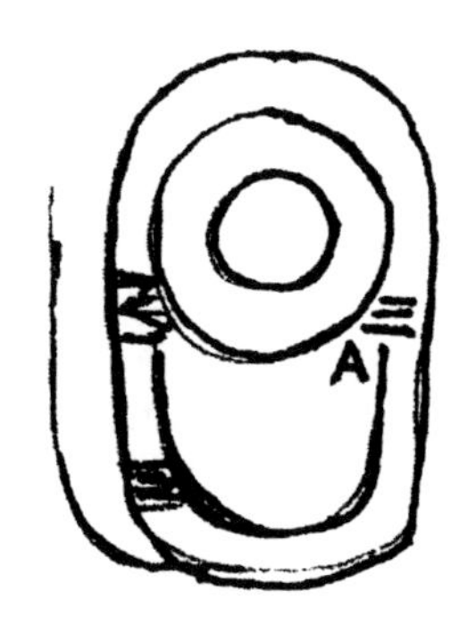

> Old Wilcox went w. us – said at A was a lock where he took out a piece of timber as deep as his shoulder, 20 ft. long: other timbers was left there: all the orchard full of rubble & stones & the moat itself w. stone on the orchard side.

In his diary, Thomas added a sketch of the tump and surrounding moat.[3]

As he grew older he continued this interest in prehistoric burial sites, archaeological remains and church monuments, and wrote notes and made drawings about his visits in his commonplace books. In 1845 he climbed Lansdowne Hill, near Bath, using an Ordnance Survey map to look for a barrow. He made notes on Roman remains at Bath, and included a sketch. Another time he purchased 'a coin of Probus found 200 or 300 yards below the site of the Roman villa at Whitchurch'.

One burial site which interested Thomas was Twyn-y-beddau. This small mound, beside the mountain road underneath Hay Bluff at a height of some 1,080 feet, was visited by the Woolhope Club from Hereford in May 1871, and an excavation of the tomb commenced.[4] Thomas recalled a visit he made some four years earlier with another clergyman when they measured the height of the tump with the aid of an alpenstock and umbrella – such accoutrements no doubt always being carried by the well-prepared walker in the Welsh hills! Webb wrote to his young friend and colleague Kilvert, encouraging him to visit the site, and including a small sectional drawing of the tump. Kilvert did so, and records in his diary the results of the excavation and the successful hoax that was perpetrated on the excavators.[5]

Webb did not join the Woolhope Club, but was elected an honorary member in 1875. Club transactions include his letter written when members were to undertake a field visit in 1884 to the Little Doward, a hill fort near Ross-on-Wye. He quotes extracts from a notebook he had kept in 1850 whilst at Ganarew, where he was able to look over the Wye valley to the hill. Much of the letter is taken up with information about the damage done to the fortifications by the newly rich ironmaster who had purchased the land and carved out carriageways and built an iron observation tower for the delight of his family and to impress his visitors.[6] However, Webb had written an exact description of the remains of the hill fort before the damage was inflicted. He also quoted local stories about the finds of burials and prehistoric animal bones in the nearby (King) Arthur's cave, and cited a passage from the *History of Geoffrey of Monmouth*, discussing the meaning of the Welsh names quoted in the Latin text.

For a man of the cloth, it may be surprising to note that Thomas was interested in tales of military endeavours. In 1833 when invited to dinner at Treago house – the principal residence in St Weonards – in the week that he became curate at the church, he listened avidly to the tales of the Peninsular Wars. A guest, Colonel Oates of the 88th Regiment of Foot, was obviously a great admirer of the Duke of Wellington who, he said, always ordered his infantry officers to ride horses on the line of march so that they might be able to take care of their men at the end of the march. His predecessor Sir John Moore apparently insisted that the officers should march alongside the men, 'whence they were as much knocked up as the men, & utterly unable to provide for their wants', a difference of opinion which provoked discussion and was carefully recorded in Thomas's diary.[7]

As a curate, Webb collected local stories and traditions whilst travelling in the area and by talking to his parishioners – especially the older members of his flock, whom he encouraged to talk about past events. He was developing a lifelong interest in oral reminiscence. There were stories of murder and wife-beating, tales of superstition and ghosts. The fear of witchcraft was by no means dead, even amongst younger people, and Thomas half-believed in protective devices against evil that had changed little from medieval times.[8] He wrote down most of these tales at length, including conversations that were held with the 'great witch', the devil, the ghostly apparition, taking at face value the experiences of those on 'whose veracity' he could place 'the most unquestionable dependence', though he did not attempt any explanation. Common sense made him add, however, that in his opinion some stories owed more to drink than to the devil, or to cunning, as when an employer feigned stories of ghosts in the cellar to prevent his servants from stealing his wine. Thomas also debated, as in this instance, the probability of truth:

> 1834 November 24: the standing stone (a druidical monument, apparently standing erect in the bank of the road between Hongar Hill & St. Weonards) is a place very much dreaded from the following circumstance. A man had stolen a sheep, & having tied the hind legs together, put them over his head to carry it. He set it down to rest on the top of the standing stone, & the sheep slipping down behind the stone, the legs

> pressed so upon his throat as to choke him. This is very probably a true story, as not at all likely to be invented. & the more so from a circumstance now not agreeing with the state of the stone, which also proves the event to be of great antiquity. The stone is now so embedded in a lofty bank with a hedge upon it that nothing could slip down behind it; but no doubt it once stood insolated like a Maen Hir.[9]

Thomas recorded oral history throughout his life and this activity took a more organised form in the last years of his father's life when he encouraged John Webb to reminisce about events and the people he had met in time past. These stories Thomas wrote down in two notebooks over a period of four years from 1864 until two months before his father's death.[10] Reading them at the beginning of the twenty-first century one is carried back to the night that the army of Bonnie Prince Charlie rested at Derby in 1715 (a story told to John by an old lady he met on a coach, whose mother had entertained the Prince's officers), to London where he watched street entertainers as a child, to a recounted incident in the Peninsular War or to his own memories of meeting with Mendelssohn. In a few of the stories Thomas adds in parenthesis his own memories of the occasion. One such case is where John remembers seeing the bodies of those executed for murder hanging by the road side as a deterrent to others. Thomas adds that an uncle had taken him to see such an exhibit in London – no doubt a suitable educational opportunity for a youthful nephew. We would agree with John's opinion that 'it was a dangerous and disgusting practice' – yes, but not only because: 'it frightened the horses'. One such site, visible from Hardwicke Vicarage, was known as Hardwick Green or Gibbet Green – names which are still in use.

Thomas appreciated, as many do not, that the changes viewed in their own lifetime quickly become of historic interest. For example, methods, speed and ease of travel changed rapidly during Thomas's life. Road travel when Thomas was a little child could be a hazardous undertaking. He recollected and recorded that his father and mother whilst travelling from Ross to Gloucester, were not prepared to set out until the hired chaise was pulled by four horses, rather than the normal two, to be certain of getting through the mud that would be encountered on the major road linking London to south and west Wales. Thomas used the train and occasional steamship journeys as these became available and convenient and commented upon the difference with approval and enthusiasm.

Over the years John Webb had made extensive visits to the British Museum and the Bodleian Library and he had had access to the libraries and muniment rooms of many of the principal landowners in

Revd John Webb (date of photograph unknown).

Herefordshire, with whom he was in frequent correspondence. One of his great interests was the English Civil War 1642–46 as it affected the county, and in some of the principal characters of that period. When he retired to Hardwicke in 1860 he brought with him a large library of books, collections of original and copied documents, and bundles of notes,[11] and he continued to gather information about the history of his county.

It was after John Webb's death in January 1869 that Thomas became a thorough but perhaps reluctant historian, for he dutifully began the task of bringing order to his father's extensive collection so that it could be completed and published. So, in 1870 Thomas wrote to Ranyard that the Camden Society had agreed that he could edit 'the Old Colonel, so that must go forward speedily "worse luck" as the country people say, though I ought not to say so, as a tribute to my dear father's memory'.[12]

The 'Old Colonel' was John Birch – a successful Parliamentarian regimental commander and later an equally successful Royalist business man, financier and Member of Parliament after the Restoration in 1660. A *Military Memoir* of Birch had been written by John Roe, the

Colonel's secretary and quartermaster. This was given to John by a distant descendant of Birch. At the age of ninety-two John studied the *Memoir*, and added more than two hundred pages of notes! In April 1871, when Thomas wrote to Ranyard expressing surprise that a second edition of *Celestial Objects* was wanted, he added: 'an annoyance just now with that Birch rod over my shoulder'; and by August 1872 he was writing: 'I hope soon to rake Col. Birch and his regiment out of the box where they have lain so long. Most Happy I shall be to rout them all.'[13] At the beginning of 1873 he was faced with reading the proofs of the *Memoir* and *Celestial Objects*:

> The old covetous canting colonel has also made considerable progress – more than 100pp printed but I have found the corrections of both at once a heavy grind for eye and brain – and shall be ready to jump for joy when these two tasks are completed.[14]

His labours were not ended until a number of queries on parts of the story were resolved with the help of Ranyard, the index created and final proofs checked. Finally, the Camden Society published it:[15]

> The Colonel has just arrived from London by train. He has not yet taken off his overcoat, but seems safe and sound. I hope to introduce him to my wife presently.[16]

The volume included a preliminary notice by Thomas, stating that it was

> ... the production of an author, the greater part of whose ninety third year was employed in the preparation ... [and] was at the time of its interruption nearly fulfilled: the latter part of the manuscript had not received his final touch, and the close was left very imperfect. These deficiencies could be ill supplied even by one who best knew his revered parent's intentions ... [but] he has done it to the best of his ability.

The notes added by John and edited by Thomas included information gained from an earlier writer whose researches had led him to include information hostile to Birch's record in Herefordshire. The Webbs accepted these comments uncritically. Birch, as a Parliamentarian officer, had threatened the life of the Dean of Hereford Cathedral when he captured the city in 1645, and had also made serious speculations in church property.[17] The aged John, as a Royalist churchman, would not look kindly on such behaviour, and Thomas, being of the same persuasion, edited his father's work without redressing the balance.

Generally, as Geoffrey Aylmer states (below), the Webbs were not unduly partisan in their published opinions. However, it would seem that they were not proud of their Puritan ancestry. In an aside in a letter of thanks to Arthur Ranyard, who had supplied some information on troop movements in the Civil War, Thomas wrote:

> 24 Feb. 1873: The Rushworth matter was exactly what I wanted. Regt. was only expected at Chester instead of having been there. The Col. Webb I am afraid was one of our family but we have – as I believe – a full counterpart in that gallant Col. Webb who was amongst the last cavalry officers that held out for the Crown.[18]

Once the *Memoir* of Colonel Birch was published, Thomas felt obliged to turn his attention to the partly finished work *Memorials of the Civil War in Herefordshire*. John had delivered a paper to the Society of Antiquaries as far back as March 1836, after 'long years of thoughtful study', and the subject continued to form 'the cherished pursuit of a great part of middle and advancing life'.[19] Thomas grappled with the problems of completing the work – checking references, sketching out material in the appropriate place in the narrative and, indeed, finishing a major part of the text.

In a series of letters to Mrs Acton Scott, of the eponymous estate in Shropshire, to seek detailed information from this leading local family, Thomas provides some insight into the task he had set himself (and Henrietta):

> The book progresses – as usual slow and sometimes very troublesome – for instance where I have to fill up blanks and have no idea where to look for information. My wife and I were both engaged in Civil War ('tho domestic peace) last night – I as usual verifying references – she examining loose papers ... It may take hours to verify some questionable references – and there are quantities of them. I have so few books of reference here ... defeated in the British Museum when last in town for my health ... some days of London fog when reading would have been impracticable ... Fortunately I have a cousin's daughter in London who honourably ekes out a very slender means by copying in the Museum and who can to some extent supply the deficiency.[20]

These concerns are vividly set out in the preface which Thomas completed in March 1879. He wrote that the prepared manuscript was only 'a fragment' and certainly the footnotes initialled TWW increase considerably in number towards the end of the first volume. Then he admits that after page 183 of volume 2:

> At this period of the history, the pen which had so diligently traced the progress of the conflict was finally laid aside; nor is there any indication of the mode in which a few unconnected and revised fragments would have been interwoven in the future narrative. They have, however, been as far as possible incorporated in the following attempt to continue a story which could not well be 'left half told'! Where introduced as notes, such material will be distinguished by the initials JW.

Consequently the work involved to get the two-volume history to publication with Longmans must have been considerable. It is a readable book with a narrative of pace and vigour. Footnotes are copious but add fascinating and sometimes amusing or ironic details that bring vivid life to the narrative. In a footnote to John Webb's spirited description of a skirmish at Powick, Worcester, Thomas adds that Vicars (a contemporary writer), 'who seems not very accurate in his account of this (or any other?) affair', said that the Parliamentarians

> ... stopped to sing a psalm before they advanced; which is not very probable (though frequently their practice) because, unless the wind had carried the voices of 500 men in an opposite direction, they were so near that [Prince] Rupert's party must have heard them.[21]

The Webbs also prepared illustrations for the book. Thomas searched diligently for suitable portraits of the principle protagonists and the text was enlivened with sketches of relevant places. Some of these are

Brampton Bryan castle. Wood engraving after a drawing by Henrietta Webb.

attributed to John Webb and others, but most are anonymous. It is likely that they were executed by Thomas and Henrietta, since in a letter to Ranyard Thomas spoke of his wife making 'a drawing of Brampton Bryan castle to be cut in wood'[22] for the book and the other drawings are in only two styles. The gathering of photographs involved many letters and visits to houses in the county in order to obtain permission to photograph family portraits.

The last part of the book is much less detailed. Indeed, if it contained the same level of detail as the first 573 pages it would have needed a third volume. Maybe Thomas's devotion did not extend to completing it in that way. Besides the further time required, Thomas also had to organise the publishing. It would seem that he probably paid for this himself. He tried to get up a subscription list but this had a very poor initial response and it would seem that he abandoned it.[23] He wrote somewhat worriedly to Ranyard about news of a rival book 'by a real faithful antiquary & historian – or merely a charlatan – a penny a liner on stilts', and averred that if he could afford it he would illustrate 'pretty freely with photography – thus mounting heavier artillery than my antagonist – a well illustrated book after all has a better chance'.[24]

Thomas was not free of professional rivalry. His comments on the 'rival book' were scathing:[25]

> I see from the 2nd Vol. easily enough that Phillips is no formidable ~~antagonist~~ rival. After all the flourish of trumpets in his prospectus, I do not see that he has ever consulted the Rupert correspondence or the Brereton despatches – and I should have a poor opinion of his learning acuteness from one little fact – You may possibly have noticed that he is 'brought up all standing' as the sailors say, by the expression cone grew in one of his documents, which he has enclosed in guillimets and put (sic) after it. I should have thought there was little difficulty in seeing here the 'coneygre' of our ancestors ... Whether a sight of his first Vol. may tend to make me think worse or better of him I don't know but I hardly expect the latter.[26]

For most of a decade therefore, much of Thomas's – and some of Henrietta's – leisure time was devoted to these two historical works, showing not only the filial respect Thomas held for his father but also the lengthy and meticulous attention to detail that was applied to whatever task he undertook. Perhaps comments by two modern historians best sum up the historical achievement of the Webbs. Ronald Hutton writes that there was a 'charm of style' and a 'use of sources long vanished'. He continues by saying that their work remains 'an exemplar of both these qualities'.[27] And Geoffrey Aylmer, pointing out their lack of basic research materials, stated:

> The two Webbs' achievement was the more remarkable in that, while both obviously felt passionately about the issues which had divided the country in the times of Charles I and Cromwell, neither carried royalist-anglican partisanship to the point of overlooking faults and misdeeds on either side, blurring the historical record, or of mistaking the evils of civil war – which it takes two to make as well as to resolve. Another source of strength was their obvious intimate knowledge of the county's landscape and topography, which I can only suppose came from having ridden over it on horseback ... true topographical 'feel' cannot be achieved merely by driving around in a car. Moreover, for clergymen rather than soldiers, the two Webbs' excellent treatment of military movements and encounters seem likeliest to be explained in this way.[28]

If Thomas had not made his name as a stargazer and populariser of astronomy he might well have earned a niche as a thorough and lively local historian.

Notes and references

1 T.W. Webb, Acton-Scott letters, MS in Hereford Record Office.
2 Placed in Tretire church by the Webbs, this may be seen today in the now redundant church of Michaelchurch. This simple building stands in a circular churchyard, where there is a sense of the numinous which long precedes the Christian era.
3 Diary, 3 and 7 November 1826 and 17 December 1829.
4 Woolhope Naturalists' Field Club – the Herefordshire society devoted to 'the practical study, in all its branches, of the Natural History and Archaeology of Herefordshire' – was founded in 1851.
5 Kilvert, *Diary*, 26 May to 2 June 1871.
6 In turn, the remains of these philistine actions are of interest to today's archaeologists.
7 Diary, 18 July 1833.
8 See C. Phythian-Adams, 'Rural Culture', in G.E. Mingay (ed.), *The Victorian Countryside*, 2 vols. (Routledge & Kegan Paul, 1981).
9 Diary. 'Maen Hir' is Welsh for 'long stone', set upright in the ground. They are usually ancient markers of a topographical or ritualistic nature.
10 J. Webb and T.W. Webb, 'Anecdotes', MS in Hereford City Library.
11 That professional historians consider some of the transcriptions to be of lost originals can be evidenced from R. Hutton, *The Royalist War Effort*, second edition (Routledge, 2003) p. 244.
12 Webb to Ranyard, 15 January 1870.
13 Webb to Ranyard, 26 August 1872.
14 Webb to Ranyard, 28 January 1873.
15 J. Webb and T.W. Webb (eds.) *Military Memoir of Colonel John Birch* (London, Camden Society, New Series, Vol. VII, 1873).
16 Webb to Ranyard 11 December 1873.

17 E. Heath-Agnew, *Roundhead to Royalist: a Biography of Colonel John Birch, 1615–1691* (Express Logic, Hereford, 1977).
18 The 'gallant colonel' was of the direct line of the Webbs of Odstock; see Chapter 1, n. 7.
19 T.W. Webb, Preface to John Webb's *Memorials of the Civil War in Herefordshire*, ed. T.W. Webb, 2 vols. (Longman, Green & Co. 1879). This work is subsequently referred to as *Memorials*.
20 T.W. Webb (n. 1)
21 *Memorials*, **1**, p. 145.
22 Webb to Ranyard, 4 January 1878.
23 The author is indebted to Adrian Harvey for this information.
24 Webb to Ranyard, 6 May 1874.
25 J. R. Phillips, *Memoirs of the Civil War in Wales and the Marches, 1642–1649* (Longmans, Green & Co., 1874).
26 Webb (n. 1).
27 R. Hutton, Introduction, in *The Royalist War Effort* (n. 11).
28 G. Aylmer, 'Who was ruling in Herefordshire from 1645–1661?', paper presented to the Woolhope Naturalists' Field Club, 1972.

Chapter 6

Webb's observations of earthquakes

Roger M.W. Musson

Then I became impressed with the fact that for the first time in my life I had experienced in full consciousness one of those wonderful phenomena – according to my earnest wish.[1]

Amongst Webb's papers is a small soft-backed notebook measuring approximately 7 x 4¾ inches. On the cover is written:

Miscellaneous Observations
by
TW Webb
1874
NB There are more note worthy observations
in this little book than might be inferred
from its appearance

This book contains jottings made between 1874 and 1883 on various natural phenomena, such as unusual meteorological conditions. Amongst these are a large number of reports of earthquake vibrations being felt by Webb and his wife at Hardwicke vicarage. These latter are of particular interest today because of the importance of studying historical earthquakes in order to be able to estimate future hazard.

These notes have apparently caused some ribald amusement to those who have consulted the Webb manuscripts at various times, because many of them take the form of comments such as: 'My wife and I felt another vibration in bed'. Some modern readers – conditioned, perhaps, by the twentieth-century myth that earthquakes do not occur in Britain – have dismissed these observations as fanciful or eccentric. In fact, they show Webb to have been a meticulous observer. As the following will show, all can be taken at face value.

Transcript of the earthquake observations

In the following summary of the notebook, all the observations are listed in the order that they appear (which is not strictly chronological, as Webb occasionally jotted things down on scraps of paper and entered them later). Those of seismological interest are given in full with the original punctuation preserved, while meteorological and other notes are merely indicated by topic. The illustration on p. 95 shows the location of the various places mentioned in the transcript and the subsequent discussion. The transcript is followed by a table summarising all the earthquake observations, including those few mentioned by Webb but not actually felt by him.

1874

March 24 – Solar halo.
May 6 – Diffuse light in sky.
May 7 – Strong light amongst dark clouds.
May 21 – Solar halo.
June 6d. 12h 10m (after midnight). Being quite awake I felt a vibration in the bed, & mentioned it to my wife – she had not noticed it, being more sleepy, but then said she perceived it, & that it was from right to left, which however I could not confirm.

1875

Dec. 17d. 7h. 46m. About half a minute previously, there had been a slight sound at the W. window of the drawing room of Hardwick Vicarage, as if it were very gently tapped, or shaken by the wind, though it was perfectly calm. This made the 2 little dogs very noisy; & at the time specified, as I was sitting with my upper arm hard pressed against a round table, I felt 3 or 4 considerable vibrations in it, & thought a dog might have run against it; but on looking at the position of the feet I do not think it could have been so. But my wife noticed nothing at her table; & there was no sound or shaking in the room. The creak or tap at the window was repeated at 8h.28m. – I saw no subsequent account of an earthquake in the newspapers.
December 22 – Meteor.

1876

Jan. 5d ('possibly at a'[erased]) in the early morning, i.e. a little after midnight, my wife & I, then lodging at 18 Rodney Terrace, Cheltenham, & being both awake, felt the bed vibrate under us for a considerable time, perhaps 20s. or 30s.
Jan. 10 & 11d. About 7m. before midnight at Hardwick Vicarage began to vibrate with a shivering motion, which might last 15 or 20s. This was repeated at 12h. 12m. & at 12h. 33m. In the 2 latter cases I looked at the curtain, but as far as the light of the nightlight would shew it was unmoved. We were both awake, & both perceived it alike.
Jan. 13d. About 1h. 30m. A.M. we both noticed the same sensation, though in rather a slighter degree.
Sept. 15–16d. Sometime between midnight & 1h. being perfectly awake, at Hardwick, I felt the bed vibrate for perhaps 1m. – My wife was asleep at the time – but I know that I was not dreaming, having changed my ring to the other hand, on purpose not to forget it in sleep.
Oct. 13d. Some time in the early morning – probably after 4h. my wife distinctly felt the bed jarred under her for a little time – I was asleep – I have omitted an instance between the last & this, when I felt the same very distinctly, but did not enter it, because she, though awake, did not perceive it.
September, date uncertain, probably 23 – strange light in sky.
August 12 – Meteor.
November 8 – Meteor.

1877

Jan. 30d. About 4h. 45m. AM. I thought the bed vibrated under me slightly but rapidly for nearly a minute – & the same was repeated in a less degree twice before day. But I [erasure: dis – insertion: cannot altogether] trust them [erasure: all], as they took place on waking, & before I was sufficiently collected to be quite certain.
February 2 – Bright zodiacal light.
May 7 – Solar halo.
Nov 5d. A long continued vibration in bed – between 11h. 30m. & 12h. – followed after a short interval by another – each perhaps for 8s. or 10s. – with a 3rd feebler following. I may have been drowsy, but I think it pretty certain. My wife seemed to be asleep & I did not disturb her.

1878

Apr. 9d. About 0h. 7m. there was a very perceptible vibration [insertion: in the bed] of rapid character – seemingly transverse – for about 40" or 50" [seconds]. I was drowsy – but very certain of it. My wife, in pain from rheumatism, said she had not been attending to it. About 0h. 40m. I believe, it was renewed in a slighter degree – she was then asleep.

April 21. [20 corrected to 21 with insertion: (21)d] At 0h. 35. A.M. we both felt, most distinctly, a vibration of the bed, seemingly transverse, & so distinct that I asked my wife to see if there was any trembling in the curtains but she could not perceive it. It probably lasted 30 or 45s.
April 25d. At 0h. 30m. another vibration of the bed felt by us both; not quite so striking or long as the last.
August 28 – Meteor.
NB. Some time in the early part of this month [August] we both felt a very distinct [erasure: rocking; insertion: vibration] of the bed in a transverse direction, lasting perhaps a quarter of a minute – which I forgot to record at the time.
Oct. 26. At about 0h. 5m. my wife awoke me to notice a vibration of the bed, which I felt very distinctly, tho' I did not remark anything as to its direction. The nightlight shewed perfect stillness in the curtains. 20m. afterwards it was repeated for a short time & we both felt it again – & later I thought it was renewed – but she did not perceive it.

1879

Mar. 1d. 0h. 45m. My wife & I being both awake, felt a very distinct quivering of the bed for about 10 or 12s.
Aug. 25d. 11h. 40m. or thereabouts, my wife told me that the bed was shaking – but I had been too sleepy to notice it. 10m. later, however, it returned with especial distinctness for perhaps 10s: both of us then being quite awake. The vibration was so strong that I looked to the curtains, & believe they might have been seen to quiver had not my sight been too short & the night light too feeble.
Oct. 19d. Lady Wyatt sitting in the drawing room at Hardwick, probably about 3h. P.M. felt a distinct vibration in her chair; & said she had frequently felt it in bed at Dimlands Castle (near Cowbridge.) She had never met with anyone who had noticed a similar phenomenon except ourselves, and her sister Miss Nicholl.
Nov. 19. Mr Espin, visiting us, said he had also felt such a vibration, & I think it was here, whither he had only come yesterday.

1880

May 13. About 0. 15 in the morning (after midnight) we both felt the bed tremble for a considerable time, & very distinctly.
May 3 – Thunderstorm and black rain.
End of Oct. or beginning of Nov. my wife & myself felt a long-continuing vibration of the bed – perhaps for 45 s. It was about 30 m. after midnight & we were both fully awake.
Dec. [erasure: 1] 9d. 0h. 15 or 20m. My wife asked me if I did not feel the tremor – but it was so slight that I was doubtful – it seemed as a mere jar – soon after however the real trembling came on, tho' slighter than usual & of short duration.

1881

Jan. 9d. About this time, I am not certain of the day, I felt a vibration repeated 3 times at intervals of about 5m. some time I think after midnight. My wife however did not perceive it. I thought also – about this time, that I noticed such a vibration in my chair during the day-time. [Inserted at bottom of page, its correct place indicated by an asterisk and a note: see bottom of page] Jan. 14 or 15. midnight. Both of us felt a long, tho' not strong bed-shake.
[Erasure: Feb. 3 or 4. My wife felt a bed vibration about midnight. I was asleep at the time.] [Above this erasure is an inserted entry, also erased: Feb. 4. Soon after midnight, early morning, my wife felt bed tremble, I was too sleepy.]
Feb. 6. evening. I felt an easy chair vibrate under me. My wife in another chair did not perceive it, & it did not affect the flame of a lamp. Towards midnight my wife felt it. I had not gone to bed.
Feb 6 – Unusual ray of light; also aurora on January 31.
May 3d. About 11h 45m P.M. we both of us felt a vibration of the bed for some little time.
May 5d. The same observation as the last, about the same hour.
June 5d. My wife felt very distinctly the chair vibrate under her for a short time about 6h. 46m. P.M. in the library at Hardwick Vicarage.
July 28–29. About midnight we both felt a vibration in the bed.

1882

May 8 – Black rain.[2]
Sep. 2d. Sometime in the early morning my wife felt a strong shaking of the bed. I was asleep.
November 15 – Zodiacal light.
November 17 – Aurora.
Nov. 28d. 11h 35m. my wife & myself both independently thought we felt a jerk or shock in the bed – without any sound, & instantly passing. We may neither of us have been very wide awake, but I think it was a fact.
Dec. 6d. About 12m. before midnight my wife & I, being both awake, independently felt a kind of shock in the bed, as if a blow had been given to the bedstead. No sound attended it. It was probably much the same thing as on Nov. 28, but we were both more fully awake on the present occasion.
December 14 – Black snow.

1883

Jan. 16. A little before 5h P.M. my Wife & Miss Frances E.J. Thomas being at tea in the drawing room at Hardwick Vicarage, they both heard a loud sound, which seemed to be in the direction of the road (S.W.) loud enough to interrupt their conversation. Miss T. asked what it was – being startled – my wife replied it was a trap – but noted afterwards its sudden ceasing – unlike a vehicle – & Miss T. replied 'Do traps shake the room?' for she felt it trembling beneath her feet. – I who was probably on the way to or from the telescope-house, heard & felt nothing, & no further notice as taken till the next day – when a note came from Mrs H. Allen, the Priory enquiring about it – giving the time, which we had not noticed, as above, & saying that she & Mrs J. Trumper & Miss Greswell, who were calling there, 'we all felt the room shake under us – & my maid who was in one of the bedrooms heard all the crockery gingle'. We found it had been heard at the Clock Mill, where Mrs Harris thought the mill had been set going, till her son came & told her there was no water. I heard afterwards that at Whitney, men ploughing perceived nothing, but the horses were much alarmed, & the men could not understand why. Mr Seacome told me turkeys had flown up into the trees. Mrs Phillott, Staunton-on-Wye, wrote that [breaks off].

(Jan. 27.) The Rev. J.D. La Touche, vicar of Stokesay, Salop, states that a distinct, though very slight shock of earthquake was felt there on the 27th ult. – Guardian, Feb. 7.

Feb. 9d. 10h. 10m. Another Earthquake. My wife & I were in our room, when a strange rattling or knocking sound, lasting perhaps 2s or 3s proceeding from a writing cabinet at which I had been sitting, surprized us both. She thought I had been knocking or hammering something – but I was standing quietly at the mantelpiece. I thought she might have shaken a board in crossing the room – but though she had done so she was sitting at the moment, nor had there ever been such a noise from the floor or in the cabinet. I suspected the cause; though we felt no shake – as to noise there was such a tempest of wind and rain that it might still have passed unnoticed. I found afterwards the noise proceeded from a medicine bottle lying on its side, & knocking against the back of the cabinet – Next morning my clerk of his own accord asked if there had been an earthquake – for after he had gone to bed, & as he thought about 10h. (but he had probably been asleep) he heard a rattle in his room. At Bredwardine Cottage where we went next day, Miss C. Newton, knowing nothing of what we had heard, said she had heard a shaking of things in her room, after 11h, when she was in bed – she thought it had awakened her – somebody else in the house heard the same.

Mar. 22–23d, about midnight, at Hardwick, we both felt a longish continued jarring of the bed.

Mar – some time in latter half, there was a description of a rattling noise heard by many persons in Leominster, which was evidently an earthquake – but the [insertion: news]paper got unfortunately destroyed. The account however was far from an intelligent one.

Apr. 25–26d. At Hardwick, between 25m & 30m past midnight, my wife was awakened by a vibration of the bed & called me. I never felt a more decided one. The pulsations were lateral, & as I should guess about [erasure: 200; insertion: 150 or 200] per minute. I could not see any movement in the curtain but the light in the room was feeble. After some

30s. the vibration became feebler, not slower, till it subsided into perfect stillness.
May 3 – Black rain.
May 6d. A slight jarring of the bed, about 11h 25m. perceived by my wife, who woke me, & I felt the close of it.
June 11d. 0h. 10m. My wife & myself were awakened, as we both believed, by a vibration in the bed, which seemed rather strong, but was not of long continuance.
Aug. 19–22. My wife & myself had not noticed any vibration for a considerable period – but about this time we both felt one in bed – but we were not fully awake.
September 11 – View of the dark side of the Moon.
November 17 – Strange afterglow [this and the following observations are doubtless due to Krakatoa, unbeknownst to Webb].
December 23 – Unusual dawn light; beautiful sunset.
December 15 – Unusual dawn light.

Summary table of earthquakes recorded by Webb

Year	*Month*	*d*	*h*	*m*	*Place*	*Comments*
1875	Dec	17	19	46	Hardwick	
	Dec	17	20	28	Hardwick	
1876	Jan	5	0	30	Cheltenham	Possibly not an earthquake, time approximate
	Jan	10	23	53	Hardwick	
	Jan	11	0	12	Hardwick	
	Jan	11	0	33	Hardwick	
	Jan	13	1	30	Hardwick	
	Sep	16	0	30	Hardwick	Time approximate
	Sep or Oct			night	Hardwick	
	Oct	13	4	30	Hardwick	
1877	Jan	30	4	45	Hardwick	
	Nov	5	23	45	Hardwick	Time approximate
	Nov	5	23	50	Hardwick	Time approximate
	Nov	5	23	55	Hardwick	Time approximate
1878	Apr	9	0	7	Hardwick	
	Apr	9	0	40	Hardwick	
	Apr	21	0	35	Hardwick	
	Apr	25	0	30	Hardwick	
	Aug				Hardwick	
	Oct	26	0	5	Hardwick	

	Oct	26	0	25	Hardwick	
1879	Mar	1	0	45	Hardwick	
	Aug	25	23	40	Hardwick	
	Aug	25	23	50	Hardwick	Time approximate
	Oct	19	15		Hardwick	
	Nov	19			Hardwick	Possibly the 18th
1880	May	13	0	15	Hardwick	
	Oct or Nov		0	30	Hardwick	
	Dec	9	0	15	Hardwick	
	Dec	9	0	20	Hardwick	
1881	Jan	9	0	30	Hardwick	Time approximate
	Jan	9	0	35	Hardwick	Time approximate
	Jan	9	0	40	Hardwick	Time approximate
	Jan	9		day	Hardwick	
	Jan	15	0	0	Hardwick	
	Feb	4	0	10	Hardwick	Time approximate; deleted entry, may be same as preceding
	Feb	6		evening	Hardwick	
	Feb	6	23	45	Hardwick	Time approximate
	May	3	23	45	Hardwick	
	May	5	23	45	Hardwick	
	Jun	5	18	46	Hardwick	
	Jul	29	0	0	Hardwick	
1882	Sep	2	1		Hardwick	Time approximate
	Nov	28	23	35	Hardwick	
	Dec	6	23	48	Hardwick	
1883	Jan	16	17	5	Hardwick	Abergavenny earthquake magnitude 3.8
	Jan	27	12	15	Stokesay	
	Feb	9	22	10	Hardwick, Bredwardine?	
	Mar	8	17	45	Leominster	
	Mar	23	0	0	Hardwick	
	Apr	26	0	25	Hardwick	
	May	6	23	25	Hardwick	
	Jun	11	0	10	Hardwick	
	Aug	19–22		night	Hardwick	

The earthquake observations in the context of UK and regional seismicity

Webb's observations may not be the sort of report of an earthquake that the layman would expect to find, but they are fairly typical of low-intensity observations of small British earthquakes. A number of conclusions can be drawn from them. Firstly, in all cases except for the 16 January 1883 event, the shaking is very weak – typically intensity 2 on the European Macroseismic Scale (EMS) with which strength of shaking is measured in the UK. The following points can be noted:

- Almost all the observations are at night. In the quiet of the night, an observer lying down on a bed (especially on an upper floor) has the chance to notice faint vibrations that would not be noticeable to anyone walking, standing, or even sitting. The majority of Webb's observations were therefore made when he was in bed, and most of the others were made at times of repose. Probably other events in the same sequence took place at other times of the day, and went unnoticed.

- Almost no-one else observed the shaking, except in a few cases – a further indication of the weakness of the shaking. In the case of the 17 December 1875 event, Webb felt the shaking and his wife did not, even though they were sitting in the same room. This is not uncommon with low-intensity observations. It appears that Webb was attentive to recording those cases when he could find confirmation that others had observed any of these events, and that he also checked the newspapers.

- In most cases the shaking had no effect on domestic objects. The curtains did not sway, the windows did not rattle, and pictures did not swing out of alignment. This is further evidence of the feebleness of the shaking.

There are two reasons for observations of very low-intensity earthquake shaking. One is the occurrence of a large or moderate event at such distance that the shaking has attenuated with distance to near imperceptibility, and the other is a local event so small as to generate little in the way of perceptible vibration. Here it is clear that for the most part we are dealing with the latter case. If Webb's observations represented the tail end of distant events, then firstly we would have other reports from nearer the epicentre, which we do not have; and secondly, it would be remarkable to have so many extreme-range observations from exactly the same place. This indicates a series of small earthquakes with a source close to Hardwicke. Such sequences – called swarms – are not unknown in the UK. The typical earthquake swarm features numerous small earthquakes over a period of years, with no single large earthquake that can be characterised as a main-shock. The most cited swarm activity in the UK is that around the village of Comrie, which was very intense in the years 1795–1801 and 1839–46 and has continued intermittently ever since.[3] At Comrie some large shocks (up to magnitude 4.8) have occurred, but at another swarm venue, Chichester, in the early nineteenth century, no shock exceeded magnitude 3.5.[4] Other small swarms have occurred in various places in the UK. In the case of that observed by Webb, the events all seem to have been of around magnitude 2 or less, judging by the lack of other observations.

The exception is the 16 January 1883 Abergavenny earthquake, which was a moderate earthquake observed over quite a wide area, which has been given an epicentre south-east of Abergavenny and a magnitude of 3.4 ML.[5] (All magnitudes are quoted using the local magnitude, ML, scale.) This is discussed further below.

The observation of 5 January 1876 must be excluded, as it was made in Cheltenham and therefore clearly had nothing to do with any source near Hardwicke. It is possible that this instance was due to some human cause. The remark attributed to Lady Wyatt on 19 October 1879 – that she had made similar observations in bed at Dimlands Castle – must also be dissociated from the Hardwicke reports. Dimlands Castle is near Cowbridge, between Cardiff and Swansea, and much too distant from Hardwicke for there to be any common cause. Due to the lack of further details it is difficult to know the cause of Lady Wyatt's observations. There are no other known reports of small events in the Cardiff area at this period.

From a regional perspective the Welsh–English border has relatively high seismicity for the UK. Hardwicke falls within a broad, arcuate zone of seismicity stretching from South Wales, through Hereford, the English Midlands, and up to the Pennines. Many of the larger British earthquakes occur within this zone,[6] and it is therefore not very surprising that earthquake observations were made at Hardwicke.

Of particular note is the 6 October 1863 Hereford earthquake, which had a magnitude of 5.2 ML and an epicentre in the Golden Valley, not far to the south-east of Hardwicke.[7] This earthquake was also felt by Webb, though he does not seem to have committed his observations to paper until nearly twenty years later, when he submitted a short article to the periodical *Knowledge*.[8] In this he describes his observations of the shock, which woke him from sleep by the violent rattling of wardrobe

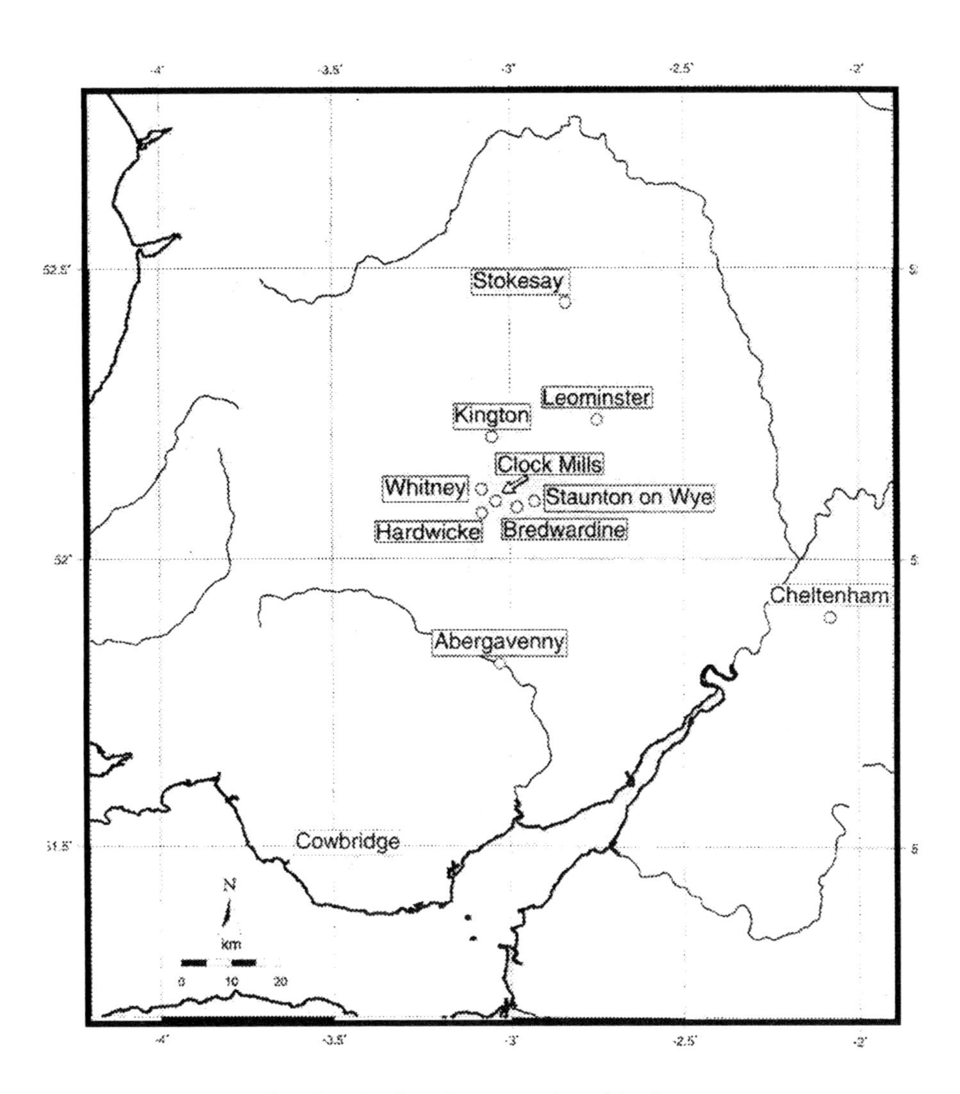

Map showing the locations mentioned in the text.

doors and pictures on the wall. A small unframed picture that had stood upright against a wall was thrown to the floor, and the chimney of an old cottage a mile distant was reported to have been cracked. Mrs Webb – in Cheltenham at the time – was also woken by the noise and shaking.

Curiously, Webb concludes his article by stating:

> Since that time we have felt two distinct shocks at my house, though of less severity. In the former instances a little dog leaped in terror from a lady's lap, though none of the party in the room felt any shock, and the sound was ascribed to a passing vehicle; a candlestick, however, was jarred on the kitchen table.

The fact that Webb should number the events felt at Hardwicke at only two, and the divergence of the description of the first of the two from any of those described in his notebook, suggests that he was perhaps referring to other events felt at Hardwicke, which he knew had been confirmed by newspaper reports. The first could have been the 30 October 1868 Neath earthquake, which was felt very widely. It was certainly felt quite noticeably in Hay-on-Wye and other places in the vicinity.[9] However, this earthquake happened at 22h 35m. It would not be impossible for people to be sitting together with a dog at this time of night, but one hesitates to make a firm identification of Webb's description with the 1868 shock. There are no other obvious candidate events between 1863 and 1882 – when Webb's article was written – besides those described in the notebook, none of which are particularly 'distinct'.

Contemporary confirmatory evidence

It would be pointless to attempt to find other data for most of Webb's observations, since it is clear from his account that he and his wife were practically the only people to notice these events, which were not reported in the local newspapers. The exception is some of the events in 1883.

As mentioned above, the 16 January 1883 earthquake is known from other sources. In addition, the 9 February 1883 event seems to have been the strongest of the local observations, being apparently perceptible also in Bredwardine, 7 km to the east. There are also two events in 1883 which Webb mentioned as being reported in the newspapers but which he did not observe personally: an event on 27 January 1883 at Stokesay; and one on some unspecified date in March, at Leominster.

The principal source of original data on British earthquakes of this period is local newspapers, which were far more copious in terms of information on local happenings than are their modern descendants.[10] The main repository of these is the British Library Newspaper Collection at Colindale, North London. Unfortunately, the physical quality of many of these old papers has deteriorated markedly in the last twenty years, and many can no longer be consulted. This is, apparently, especially true of the year 1883. According to BL staff, approximately 40 per cent of all volumes for this year are unfit for use, and three of the newspapers closest to Hardwicke – the *Abergavenny Chronicle*, *Hereford Times* and *Kington Gazette*, are all unavailable.

The opportunity was taken to look more closely at the 16 January 1883 earthquake, and data were collected from several newspapers still available and not previously consulted (*Hereford Journal*, 20 January, p. 8; *Merthyr Express*, 20 January, p. 5; *Shropshire Guardian*, 27 January, p. 5; and *Monmouthshire Beacon*, 20 January, pp. 5 and 8). To these were added the data from Webb, and other material previously known (*Abergavenny Chronicle*, 19 December 1896; *Bygones*, vol. 6, p. 177; *Hardwicke's Science Gossip*, vol. 19, p. 93, *Nature*, 25 January, p. 293; and *The Times*, 17 January) to produce a more comprehensive picture of this earthquake than previously available. On the intensity map (p. 99), numbers are EMS intensities (3 = weak, felt by few; 4 = felt by many, windows and doors rattle; 5 = strong, some objects upset; F = felt, intensity uncertain; F? = uncertain report). It now appears that this earthquake was stronger than previously recognised. The estimated magnitude, using the formulae from Musson,[11] is 3.8 ML rather than the previous value of 3.4 ML, and the epicentre is near Longtown in the Black Mountains. The earthquake was therefore very similar to the event of 27 March 1853,[12] which had almost exactly the same epicentre and magnitude. A search was made for newspaper reports of the 9 February 1883 event, but none could be found.

The Stokesay event is attributed by Webb to a report in the *Guardian* of 7 February – but which *Guardian* is uncertain. If it was the *Shropshire Guardian*, the date is in error, as the report was published on 3 February. The sole basis for this event is an observation made by the vicar of Stokesay, which he sent in a letter to the *Daily News*, from which it was copied by the *Shropshire Guardian* (3 February 1883, p. 5). There is no information on this event other than this single report.

The earthquake at Leominster can be found in the *Hereford Journal* (17 March 1883, p. 5), in which it is described as having occurred 'on Thursday in last week' – presumably indicating 8 March as the day of the earthquake. The effects in Leominster are described in some detail (giving an intensity of 4 EMS), but there is no indication that it was felt anywhere else. It seems, therefore, to have been a small event with an epicentre fairly close to Leominster.

Similar modern evidence

Any doubt as to whether such a small earthquake swarm near Hardwicke were credible can be completely dispelled by the fact that a similar swarm has occurred in recent years, and can be certified from instrumental data. This swarm was shorter in duration than the activity in the 1870s and 1880s, and the events in it seem to have been smaller, as none of them were reported as felt. Details are taken from Ritchie and Wright.[13]

The first event occurred on 30 July 1988 and had a magnitude of 1.0 ML. The last event occurred thirty-seven days later, on 4 September 1988, and had a magnitude of 0.0 ML[14] – the smallest magnitude event detected. The largest event, on 11 August 1988, was only 1.1 ML. Thus there was no clear main shock, typical of true swarm behaviour, and the total number of recorded shocks was seventeen.[15]

These events were located about 5 km south of Hardwicke. It appeared at first that the instrumentally determined epicentres formed a well-defined lineation oriented roughly north-north-east, such as might be expected from events occurring along a fault with this azimuth. However, subsequent analysis showed this lineation to be an artefact of the distribution of recording stations. Uncertainties in the epicentral locations were typically around ±2–4 km, but preferentially in a north-north-east or south-south-west direction. Focal depths were typically around 10–12 km.

It is also interesting to note that the instrumental epicentre of the 24 August 1975 Hereford earthquake (3.5 ML) was only 5 km east of Hardwicke. However, due to the limited number of stations (none close) available at that date to record the event, the location is in some doubt. Macroseismic evidence suggests a more easterly epicentre. There were no felt reports obtained from Hay-on-Wye or anywhere around it, although this may be partly an artefact of the way in which the macroseismic data were gathered.

Conclusions

The earthquake observations made by Webb appear to be methodical and accurate records of a small earthquake swarm located close to Hardwicke, and not some fantasy, as had apparently previously been supposed by some readers. The largest earthquake noticed by Webb – that of 16 January 1883 – is well documented in other sources. That the area around Hardwicke is occasionally subject to small earthquake swarms is proved by instrumental evidence of the small swarm near Hay-on-Wye in 1988, in which seventeen small events ranging in size from 0.0 ML to 1.1 ML were detected over a period of thirty-seven days.

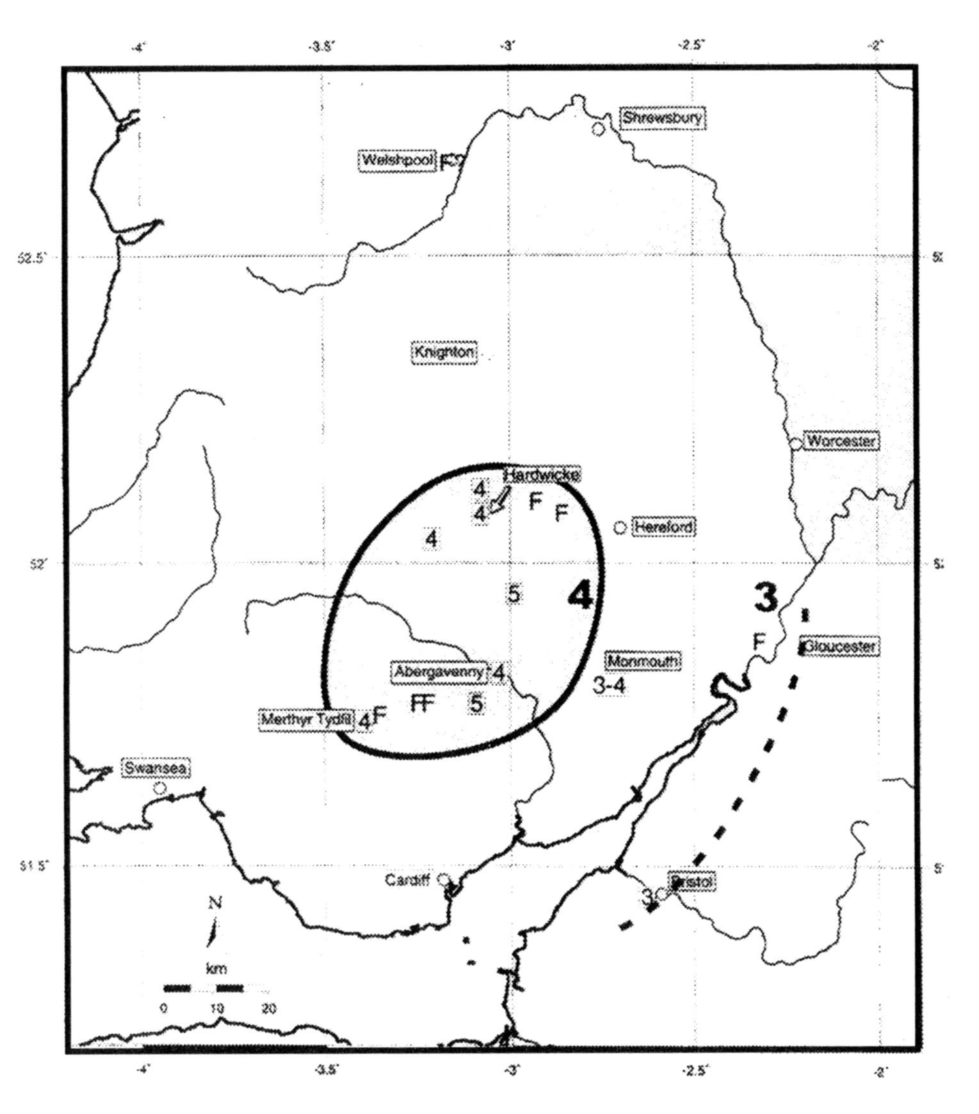

Intensity map of the Abergavenny earthquake of 16 January 1883. Numbers are intensities using the European Macroseismic Scale; F indicates 'felt' (intensity not known) and F? indicates uncertain felt report. The solid line represents the isoseismal for intensity 4, the dashed line a partial isoseismal for intensity 3.

Acknowledgements

This chapter was supported by the Natural Environment Research Council (NERC), and is published with the permission of the Executive Director of the British Geological Survey (BGS).

Notes and references

1 T.W. Webb, 'Miscellaneous Observations', MS in Hereford City Library.
2 In 1882 Webb collected some 'black rain' and sent it to George With in Hereford for analysis. The conclusion was that it was contaminated with soot – presumably from the industrial areas of the Black Country or South Wales.
3 R.M.W. Musson, 'Comrie: a historical Scottish earthquake swarm and its place in the history of seismology', *Terra Nova*, **5** (1993), 477–80.
4 G. Neilson, R.M.W. Musson and P.W. Burton, 'Macroseismic reports on historical British earthquakes. VI: The South and Southwest of England', *BGS Global Seismology Report No. 231* (1984).
5 R.M.W. Musson, 'A catalogue of British earthquakes', *BGS Global Seismology Report No. WL/94/04* (1994).
6 R.M.W. Musson, 'The seismicity of the British Isles', *Annali di Geofisica*, **39**, no. 3 (1996), 463–69.
7 R. Prowse, 'Reinvestigation of the 1863 Hereford earthquake', unpublished report to the Nuffield Science Foundation (1998).
8 T.W. Webb, 'Earthquakes in Herefordshire', *Knowledge* (15 December 1882), 469–70.
9 R.M.W. Musson, G. Neilson and P.W.Burton, 'Macroseismic reports on historical British earthquakes. VIII: South Wales', *BGS Global Seismology Report No. 233*, 2 vols. (1984); and 'Macroseismic reports on historical British earthquakes. VI: The South and Southwest of England', *BGS Global Seismology Report No. 231* (1984).
10 R.M.W. Musson, 'The use of newspaper data in historical earthquake studies', *Disasters*, **10**, 217–23.
11 R.M.W. Musson, 'Determination of parameters for historical British earthquakes', *Annali di Geofisica*, **39**, no. 5 (1996), 1041–8.
12 Musson (n. 5); Prowse (n. 7).
13 M.E.A. Ritchie and F. Wright, 'The 1988 Hay-on-Wye earthquake sequence', *BGS Technical Report No. WL/91/31* (1991).
14 Where the network is sufficiently dense, earthquakes can be detected down to –3.0 ML. Magnitude 0.0 ML is equivalent to the explosion of about 1 lb of TNT.
15 Some of these were felt by the then occupants of Webb's vicarage – Mark and Janet Robinson – but were not recorded with Webb's accuracy.

Chapter 7

Clerical astronomers

Allan Chapman

The Spangled Heavens, a shining frame,
Their Great Original proclaim
Joseph Addison (1672–1719)

Thomas William Webb was a member of a professional group which, across its denominational range, probably contributed more serious workers to the nineteenth-century astronomical community than any other. Webb was a 'Clerk in Holy Orders', a Priest of the Church of England, and incumbent of the parish church of the Holy Trinity at Hardwicke, Herefordshire. While as an Anglican he belonged to the best-endowed and most numerous clerical group in Britain, it must not be forgotten that Methodist and Dissenting Ministers and Roman Catholic priests also formed the wider body of astronomical clergymen, especially during the Victorian period after 1837. It is remarkable to note that, in this age of sometimes acute interdenominational rivalries, astronomy could bring men together from across the ecclesiastical board, as Catholics and Dissenters joined their Anglican brothers at meetings of the Royal Astronomical Society, in much the same way as Jeremiah Horrocks and William Crabtree (both Anglicans) formed warm and fruitful astronomical friendships with William Gascoigne and Richard Towneley (both Catholics) in seventeenth-century Lancashire.

But why were the Anglicans to be so prominent, and why did they control the lion's share of ecclesiastical power and patronage? The Church of England had become the 'Church Established' following King Henry VIII's Protestant Reformation of the sixteenth century, and this status had been confirmed at the Royalist Restoration in 1660. The clergy of the Church of England, moreover, occupied a pre-eminent status within the British constitution, for the Church's parish structure covered the whole of England, Wales and Ireland (though not Scotland), while in each of its regions there was a Bishop, Dean, Archdeacon, and Corporation of Canons who served a Cathedral Church and its surrounding diocese. The twenty-six senior Bishops sat in the House of Lords (and still do), and parish clergy often sat on the local magistrates' Bench (though Webb himself was not a magistrate), so that being a 'Clerk in Holy Orders' conferred an impeccable gentlemanly rank upon its holder. As will be obvious from the above, the vast majority of Anglican clerics came from comfortably-off families whose sons received a public or good grammar school education almost as a matter of course, and who invariably sported Oxford or Cambridge degrees after their names. No author has provided a more intimate or amusing insight into this world than did Anthony Trollope in his *Barchester* series of novels, written in the 1850s.[1]

Methodists, Congregationalists, English Presbyterians and other 'Dissenting' groups enjoyed full legal toleration and freedom to go about their business from the late seventeenth century onwards, for they were all loyal to the Protestant cause and not suspected of being in league with the Vatican. On the other hand, they enjoyed no constitutional patronage, and members of these religious denominations had to make their own money from business, medicine, or law, or else their very modestly funded ministry had to be paid for directly by the local congregation. Without Anglicanism's landed parish structure, the Dissenting Churches (often at loggerheads with each other on doctrinal grounds) lacked those landed endowments whose rentals paid the stipends, built the houses and maintained the churches of their Anglican brethren. And as a consequence, most Dissenting churches were hard up, and often attracted men to their ministry who were not automatically deemed 'gentlemen'. There were notable exceptions, however: William Rutter Dawes, who became the foremost double-star observer of the mid-nineteenth century, was an ex-Charterhouse public school convert to Congregationalism, who earned his living practising medicine at Ormskirk, Lancashire, and also acted as the local Congregationalist minister.[2]

But it was the Roman Catholics, or 'Papists', who aroused the greatest mistrust, for popular legend had it that they – like their ritually-burned co-religionist, Guy Fawkes – were in secret alliance with Rome against English liberties, in spite of the ancient and distinguished pedigrees of many Catholic families such as the scientific Towneleys, Sherburnes and Watertons.[3] While active Catholic persecution had long since ceased, and English Catholics were at last relieved of their 250 years of civil disabilities by the Catholic Emancipation Act of 1829, suspicion still ran deep. The Revd Canon Temple Chevallier, for instance, first Professor of Astronomy at Durham University in 1839, always suspected devious machinations from the new Jesuit seminary at Ushaw, near his own parish at Esh, County Durham, in spite of the Jesuits' own active interests in science; while Chevallier's clerical friend, Dr Gilly, believed that the Durham Jesuits were secretly plotting to 'blow up the [Anglican] University'.[4] Indeed, many people, even if they were less mistrustful than Dr Gilly, still felt uneasy about the intentions of the urbane and polished Italian and English Jesuits who staffed the new Stonyhurst College Observatory, Lancashire, from 1838 onwards, in spite of Father Stephen Perry's election to a Fellowship of the Royal Society in 1874.[5]

Until the early 1870s the Church of England effectively controlled access to higher education. Until that date, not only were the great majority of Oxford and Cambridge dons, including many science dons, in Holy Orders, but undergraduates were required to be Communing Anglicans as well.[6] Only University College London – the notorious 'Godless College' – waived all religious requirements at its foundation in 1827, though the new Universities of King's College, London (1831) and University College, Durham (1832) reasserted Anglican supremacy. These five were England's only universities until the big civic, secular 'Victoria' Universities of Manchester, Liverpool, Birmingham, Sheffield and so on came into being after 1880.[7]

Ireland's Trinity College, Dublin (founded in 1591) was firmly controlled by the Episcopalian Protestant Church of Ireland, which was in communion with Canterbury, and most of its dons were gentlemen of the cloth, just as at Oxbridge and Durham. And like Oxbridge, it was financed by landed ecclesiastical revenues, even if in Trinity's case many of its tenants were Roman Catholics. The Revd Dr Thomas Romney Robinson, who was a Fellow of Trinity College, Rector of Carrickmacross, Prebend of St Patrick's Cathedral, and Director of the Archbishop of Ireland's Observatory at Armagh for fifty-nine years, was an Ulster High Tory, and when the Parliamentary disestablishment of the Protestant Church of Ireland threatened to strip the observatory of its research revenues in 1869, Robinson complained that Ireland had lost a first-rate institution which until then had worked 'without costing the Nation a shilling'.[8]

Then, in 1845 the three new Queen's Colleges of Belfast, Cork and Galway were established in Ireland, followed in 1850 by Queen's University, Belfast. Wales' first institutions of higher learning, at Aberystwyth, Bangor and Cardiff, which did not have religious qualifications for entry, were not founded until the second half of the nineteenth century.[9]

England's major schools were also overtly Anglican institutions. Clerical Head, House and Form Masters were the norm in the great public schools, not to mention the leading grammar schools, and when Coggia's comet blazed across the skies in 1874, the Revd Mr Frederick Hall, MA, FRAS, a House Master at Haileybury, gave lectures on astronomy to the boys. One of them was the young Thomas Henry Espinall Compton Espin (the son of an academic theologian and dignitary of Chester Cathedral), whose imagination was fired by the comet and by the Revd Mr Hall. Espin – who after studying at Oxford would follow his father into the Church – became one of the most distinguished independently financed amateur astronomers of his generation.[10]

It was not, however, so easy to obtain a good education if one was a Dissenter or a Roman Catholic. It is true that men of devout Wesleyan Methodist background, such as the Cornish discoverer of Neptune, John Couch Adams, could be sufficiently close in spirit to Anglicanism to obtain a Cambridge Scholarship in 1839,[11] but if a boy came from a family that was more to the spiritual 'left', so to speak, he was lucky if there was, at hand, a good 'Dissenting Academy' that specialised in educating non-Anglicans. Manchester had an excellent one, and the future discoverer of Uranus's satellite Triton (1846) and friend of Dawes, the Liverpool Congregationalist brewer, William Lassell, attended the Academy in Rochdale some time around 1813.[12] For a high-quality Catholic schooling, however, before the founding of Stonyhurst College, Lancashire, in 1794 (from the English College at St Omer, France) and Downside Abbey, near Bath (from Douai, Flanders) in 1795–1814, it was necessary for Catholic parents to send their sons to continental Europe. The most famous of the English colleges in *ancien régime* Europe had been those in Rome, Douai and St Omer, but the atheistical French Revolutionaries suppressed Douai and St Omer, which led, ironically, to their educational foundations transferring to a new home, with toleration, in Protestant England.[13] Yet it was not for nothing that the English Catholic divine and convert from Anglicanism, John Henry Newman, strove but failed to see a Catholic university established in Ireland, although the Maynooth Seminary, founded in County Kildare in 1795 (with a substantial new endowment in 1846) to train Catholic clergy, did enjoy a distinguished reputation for science, particularly for electrical physics with Father Nicholas Callan,[14] but not especially for astronomy.

Scotland, however, had a set of religious rules all of its own. Here the controlling ecclesiastical group was the Presbyterian Kirk which, while deeply influential in Scottish society and enjoying considerable powers of patronage, did not exert the same level of formal control over the country's four universities – at St Andrews, Edinburgh, Glasgow and Aberdeen – as did the Anglican and Episcopalian Churches in England and Ireland. The Scottish universities had no official religious barriers, and English Dissenters, Irish Catholics and even Anglicans who could not afford the high fees of Oxford and Cambridge, mixed cheek by jowl with strict Scottish Calvinists, Welsh Methodists, and Wee Frees. Scotland's most famous clerical astronomer of the nineteenth century was Dr Thomas Dick of Dundee, who, while not involved in astronomical observational research, exerted a profound influence across the English- and Welsh-speaking worlds with his astronomical books, and, within Scotland, with his lectures.[15] Dick was one of those great scientific communicators who aimed to reach the widest possible readership, while his lectures to Scottish working men touched countless people. In the early 1880s the Coupar Angus railway porter astronomer John Robertson vividly recalled hearing the then elderly Dick's astronomical lectures delivered to working men in Montrose around 1848,[16] while the Bangor dock worker and early amateur builder of silver-on-glass reflecting telescopes, John Jones, spoke of the formative influence which a Welsh translation of Dick's *The Solar System* (*Y Dorspath Heulawg*) had exerted upon him many years before.[17] Dick's books were not just technical treatises on astronomy, but works suffused with a sense of God's glory as seen in the celestial creation. They aimed not simply to teach, but also to elevate and inspire.

An identical spirit runs through Thomas Webb's *Celestial Objects for Common Telescopes* (1859), and though Webb's book is concerned primarily with opening up the world of practical observational astronomy to amateurs of relatively modest means, and takes a less wide-ranging philosophical sweep than does Dick's, the intimate relationship existing between astronomy and religion is assumed to be the same.

Where the clergy of the Anglican along with the Welsh and Irish Protestant Episcopalian Churches enjoyed an especial advantage over all the Christian denominations, however, was not so much in their constitutional and social standing as in their wealth and security of tenure. And while the Church of England's clergy were not paid by the government (as is sometimes wrongly supposed), their parishes, academic appointments, or other 'benefices' all came from a network of independent endowments, which derived in most cases from the Church's corporate ownership of land. And though the Ecclesiastical Commissioners after 1840 had tried to even out clerical stipends, enormous gulfs between rich and poor parishes existed well into the twentieth century. Trollope's Dean of Barchester Cathedral, for instance, got £2,000 a year, while in 1868, Webb's Hardwicke was valued at a mere £150 a year. Fortunately, Thomas William had comfortable private means in addition to his stipend.[18]

Victorian Anglican clergy also enjoyed the freehold tenure of their 'benefices' for life. Mere idleness, absenteeism, neglect, chronic illness, or senility were not sufficient to remove an incumbent. It often required proven acts of gross criminality and action by the ecclesiastical courts to deprive a bad vicar of his 'parson's freehold'. This is why curates – often employed by the incumbent himself to do the parish work while he visited Italy or sank into feebleness in the rectory – were such a mainstay of the Church's ministry.[19] To some degree, curates formed a clerical underclass, but they put up with the system in the hope that one day they in turn would become beneficed parsons. However, as the supply of curates always exceeded the number of endowed benefices, many poorly-connected curates grew old in a succession of short-term curacies, as did the dedicated but elderly and frail Revd Mr Marley, who fell dead at the altar while celebrating the Eucharist for an absent incumbent, in Flora Thompson's late Victorian autobiographical narrative *Lark Rise to Candleford*.[20] From the 1840s onwards, social forces conspired to apply pressures on indolent clergy and make them serve their parishioners. When the reforming Frederick Temple (a teetotaller and an admirer of Charles Darwin) became Bishop of Exeter in 1869, his appointment was openly applauded by Roger Langdon, the village Station Master of Silverton, Devon – and an accomplished amateur astronomer and a devout High Churchman – because Temple would make the idle members of his clergy work! [21]

But the great generality of Anglican clergy *did* work, and often very hard, in their spiritual ministries, as did their colleagues in the other denominations. Webb was assiduous in his devotion to his flock, and as his friend, executor and obituarist Thomas Espin recorded, Webb would set out

> ... in the afternoon, with a knapsack laden with all kinds of little comforts for the sick, he would walk with vigorous stride up the hills to see some distant parishioner ... And then, when the cottage was reached, there was no mistaking the warmth of welcome and the smile of pleasure with which he was received. And there he would sit, the children gathering around him, and talk to his people of their everyday life, and local matters, making himself one of themselves, and imparting the sunshine of his life to theirs.[22]

Espin further recorded that Webb sought 'neither offices in societies nor Church preferment', but found his fulfilment in astronomy and in his Christian ministry.

Some modern writers, alluding to the sense of service felt by men like Webb, tend to veer towards the cynical, accusing them of being patronising and perpetuating the social divides of the day. Such attitudes, however, are peculiar to people who can only view human aspiration through the dark glasses of twentieth-century social theory and secularism. Many Victorian gentlemen – either in Holy Orders like Webb, or laymen, such as Prime Minister William Gladstone – not only had a secure and life-giving religious faith, but also felt that as men 'unto whom much had been given' much was expected when it came to doing good in the world.[23] Similarly-placed women reflected the same attitudes, as did Hester Periam Hawkins, the independently well-to-do wife of a Bedford Methodist minister, a writer and keen promoter of good causes who, at the age of seventy-four in 1921, was amongst the first group of women to be elected Fellows of the Royal Astronomical Society. (Her husband, the Revd Joshua Hawkins, was also an active amateur astronomer, and was a founder member of the British Astronomical Association in 1890.[24])

Thomas William Webb was the son of a Ross-on-Wye parson and member of a comfortably-off and influential Dorset and Wiltshire family, and no doubt always felt secure of his social position.[25] It was probably this assurance, and modestly independent means, that helped form a temperament in which worldly ambition and pursuit of power were not especially strong, and which enabled him to find a life of Christian ministry and celestial contemplation deeply fulfilling in itself, especially when that life lay embedded within an enjoyable social round. The Revd Francis Kilvert, who served his own very enjoyable curacy to the Revd Richard Venables at Clyro, in nearby Radnorshire, between 1865 and 1872, made several references to Webb in his now famous diary. However, the Webb that emerges there is not Thomas William the astronomer (while astronomy is mentioned, it was not a subject in which Kilvert had any particular interest), but Webb the popular parson, genial host and urbane gentleman.[26] While Webb may not have had worldly ambition, and may have been happy to devote himself to his friends and to his flock, we must never forget that he always lived as a gentleman, employing three live-in, plus 'living out', servants and a coachman, and running two carriages, as was considered only fitting for a man of his standing in rural Herefordshire.[27]

On the other hand, Webb had astronomical clerical colleagues who were more instinctively worldly in their ambitions. The Revd Dr William Pearson, for example, rose up from the obscurity of a

Cumberland smallholding, via Hawkshead Grammar School and Clare College, Cambridge, to become a public figure. His early gifts as a lecturer, writer and designer of planetaria, mainly in Lincoln, led to his rapid acquisition of money around 1800. Pearson became the Proprietor and Head Master of the fashionable East Sheen preparatory school at Richmond, London. At South Kilworth, as Rector, he built a magnificent private observatory in the two-storey brick and stone structure which is today occupied as an elegant dwelling house.[28]

By the 1810s, Pearson, then in his forties, moved easily in the highest echelons of British science, was Fellow of the Royal Society, and in January 1820 was instrumental in hosting a dinner for fourteen astronomical friends at the Freemasons' Tavern in Lincolns Inn Fields, London, out of which the Astronomical Society of London – later to become the Royal Astronomical Society – was founded.[29] Pearson was an astronomer, a patron of astronomy, and a collector of instruments. He was also a popular and generous clergyman, and South Kilworth church contains various adornments, memorials, and records of his benefactions to the old, sick and poor of the parish. Astronomically speaking, moreover, Pearson's interests lay at the intellectual cutting edge of the science of his day. Double stars, planetary dynamics, and the pursuit of problems in celestial mechanics were at the heart of his interests, along with the acquisition of state-of-the-art instrumental technology with which to pursue them.[30] Pearson moved in the world of 'Grand Amateur' astronomers: well-to-do gentlemen who took upon themselves the fundamental research of the day because the Greenwich and university observatories of the period were not sufficiently well-resourced to compete with the Imperial foundations of Berlin, Königsberg or Pulkowa.

As Pearson was a 'Grand Amateur' who belonged to the honourable profession of priest–schoolmaster, so the Revd Charles Pritchard came from an identical tradition, becoming a university professional astronomer at the age of sixty-two. In his schoolmastering days, Pritchard had educated the sons of Sir John Herschel at Clapham School, before his election to Oxford University's Savilian Professorship of Astronomy in 1870. Then, in Oxford he became the driving force behind the founding of the new University Observatory, and went on to undertake fundamental researches in lunar dynamics, astronomical photography, and the determination of stellar parallaxes. Yet not only did Pritchard see his scientific work and religious beliefs as complementing each other, but as a priest he was actively involved in the theological and constitutional debates taking place within contemporary Anglicanism, and after his death in 1893 a eulogy on his religious writings and scholarship was written by the Bishop of Worcester.[31]

The Jesuit Fathers on the teaching staff of Stonyhurst College, Lancashire, produced a succession of schoolmaster–priest–astronomers of the highest intellectual calibre. During the last few decades of the nineteenth century and the beginning of the twentieth, Fathers Stephen Joseph Perry, DSc, FRS, Walter Sidgreaves, and Aloysius Cortie in particular turned the superbly equipped Stonyhurst College Observatory into a world centre of excellence in astrophysics.[32] And if people are inclined to suspect that such men *really* entered the priesthood as a way into a scientific career, they should not forget that no-one made the primacy of their Christian priestly calling more clear than did Father Perry in the circumstances that surrounded his death on 27 December 1889.

Stephen Perry was the son of a well-do-do English Catholic family that had made a fortune from the manufacture of steel pen nibs. He had been sent to Douai for his schooling. Then, at the age of fifty-six, and with an international reputation as a solar physicist behind him, he headed the official British expedition to observe and photograph the total eclipse of 22 December 1888 at the notorious Isle de Salut, or Devil's Island, in French Guyana. Dysentery and fever were raging on the island at the time, yet in addition to fulfilling all of his scientific duties Father Perry insisted upon ministering to the prisoners, celebrating Mass, and preaching what turned out to be his last two sermons to them. In consequence, he also caught the fever, and while just about able to observe the eclipse, he died at sea on 27 December, soon after HMS *Comus* commenced her voyage home.[33] As Agnes Clerke, who knew him personally, wrote in her *Dictionary of National Biography* article on Perry, his priestly vocation always came first.

In addition to schoolmastering and astrophysical research, the Stonyhurst Jesuits were at the forefront of encouraging high-level amateur astronomy in their day. Father Perry, for instance, had led an expedition of Jesuit priest–astronomers, along with Miss Elizabeth Brown and her assistant from the Liverpool Astronomical Society's Solar Section, to observe the Russian eclipse of 1884. In Russia, indeed, the English amateur–professional party was hosted by the Moscow Professor of Astronomy, Theodor Bredichin, from whose estates they would have observed the eclipse had not cloud intervened.[34] At the time of his death in 1889, Father Perry was President of the Liverpool Astronomical Society; while three years later, in 1892, Father Walter Sidgreaves was to become the Inaugural President of the North-West Branch of the British Astronomical Association in Manchester – the branch that in 1903 would become the independent Manchester Astronomical Society.[35] Father Sidgreaves' Stonyhurst colleague, Father Aloysius Cortie, moreover, would continue to be active with the

Manchester Astronomical Society into the 1920s. All of these Jesuit astronomers were closely involved with the Royal Astronomical Society and, after 1890, with the British Astronomical Association, as well as the wider scientific world.

It was also appropriate that, in spite of Webb's Protestant caution regarding Roman Catholic 'mummery', his excellent $5\frac{1}{2}$-inch object-glass, made by Alvan Clark & Sons, of Boston, USA, which Webb had bought on the recommendation of William Rutter Dawes in 1859, was subsequently purchased by the Stonyhurst College Observatory in 1886, where it was put to good use.[36]

Clerical astronomers such as Pearson, Romney Robinson, Pritchard, and the Stonyhurst Jesuits were either 'Grand Amateurs' or else academic professionals, whose primary scientific concerns lay with cutting-edge astronomical research. On the other hand, Thomas W. Webb came to astronomy from a different direction. While his Oxford education, social background, and independent means fitted him perfectly for the 'Grand Amateur' world of high-level private research, his astronomical calling lay elsewhere. What most captivated Webb about the heavens was their ineffable beauty. To him they were an elevating presence and a source of delight and stimulus to prayer. And while no-one should think that Pearson, Perry, or any other astronomer was not equally motivated to higher things by the contemplation of the sky, that motivation took different men down different paths. To some, the stimulus could lead to a full-scale and life-long inquiry, along a direction of original research, to find out more and more about how God had put His universe together. To others, however, such as Webb, his clerical astronomical friends, and those people for whom he wrote, astronomy, I would argue, was more contemplative. By 1859, after all, no-one could hope to undertake fundamental research with a $3\frac{3}{4}$-inch refractor – Webb's 'common telescope' – or, by 1866, with a $9\frac{1}{3}$-inch reflector. Such instruments, rather, rendered wonderful *views* of objects better investigated, in research terms, with 24-inch-aperture refractors or 72-inch reflectors. What such telescopes *could* do, however – especially as they became cheaper and increasingly accessible to those of modest means – was open up the delights of the heavens to thousands of people.

First and foremost, Webb was an astronomer who loved the business of observation. He looked at the celestial bodies in very much the same spirit as an amateur botanist contemplated the beauty of flowers – as delights and wonders in their own right. This approach comes through best of all in his observing books, now preserved in the RAS library. With their careful penmanship, detailed sketches, and comments such as 'A most superb object, to which one always returns with wonder &

pleasure' (the globular cluster M13, on 1 May 1869), one senses that to Webb, astronomical objects were ultimately things of delight and elevation, rather than physical problems to investigate.[37] On the other hand, this sense of elevating joy should not be allowed to detract from the undoubted intellectual rigour which underlay all of his work and the thousands of careful observations that he made.

Webb may have been primarily an observer, yet some of his clerical colleagues and friends were clearly men who delighted in rolling up the sleeves of their frock coats to address themselves to all kinds of technical and mechanical problems. In the open and unregulated gentlemanly world of the Victorian Church of England, one finds vicars, rectors and canons who had passions for steam engines, railways, photography, ballooning, microscopy, firearms, and all manner of inventions.

Henry Cooper Key, for instance, was Rector of Stretton Sugwas, some four miles from the same Hereford Cathedral in which Webb came to occupy a Prebendal Stall. In 1859 Cooper Key was the first Englishman to figure a silver-on-glass mirror on a machine of his own construction (based on Lord Rosse's machine), set up in a building adjacent to his rectory.[38] Silver-on-glass mirrors had only just become feasible, and English clerical amateurs were to play a major role in perfecting the new technology and publicising it in magazines such as the *Astronomical Register* and *English Mechanic*. Cooper Key also inspired the Hereford schoolmaster, George Henry With, to perfect his skills as a mirror maker; and while With was not ordained, he was a licensed lay reader, his school was an Anglican foundation, and he had a close and friendly relationship, both spiritual and scientific, with the Diocesan parish and the Hereford Cathedral clergy.[39] (See Chapter 8.)

The Revd Edward Lyon Berthon was another luminary in the post-1860 world of the reflecting telescope and general invention. Rector of the magnificent Romsey Abbey church, Hamphire, Berthon invented and designed various astronomical devices that were particularly useful to the amateur. These included his dynamometer (for measuring the exit pupil of an eyepiece) and his 'equestrian' equatorial mount – a compact mount with a stirrup-like counterweight system, to carry telescopes with the new silver-on-glass mirrors that had suddenly rendered speculum-metal mirrors obsolete. Indeed, the comparatively lightweight glass mirrors were ideal for amateurs both to figure and to mount, and made available an efficient optical technology which would revolutionise amateur astronomy.

Berthon, like the dedicated inventor that he clearly was, also went on to describe a design for a cheap 'telescope house' or observatory to cover these instruments. Like every astute DIY enthusiast, Berthon

Revd Edward Lyon Berthon, standing outside his 'Romsey' observatory housing what was probably an 18-inch reflector on his 'equestrian' mount.

was a master of the art of using easily available bits of commercial technology to simplify things. His elegant 'telescope houses' – which could be built for as little as £8 10s, and also serve as eye-catching garden ornaments – had canvas-covered truncated domes that rotated on common sash window rollers that, by 1865, could be purchased from any ironmonger's shop. Not bothering to patent or manufacture these little buildings, he published full construction details in the October 1871 number of the *English Mechanic* – mentioning, moreover, that the Revd Mr Webb was delighted with his![40]

Besides astronomical equipment, Berthon also invented devices and equipment for the navy, the most notable and enduring being the collapsible lifeboat – a craft whose manufacture, in those pre-inflatable-dinghy days, provided a major source of employment for Romsey, at the Berthon Collapsible Life Boat Company, until at least as late as 1911. This invention saved unreckoned hundreds of lives at sea.[41]

Large-aperture, silvered-glass mirrors were also of great importance to the research work of one of the Church's most dedicated, eccentric

Revd T.H.E.C. Espin in January 1911, wearing his winter clothing and with his cat Kip.

and colourful clergymen: the above-mentioned Thomas Henry Espinall Compton Espin. Using a 17½-inch-aperture reflector by Calver (and a 24-inch in 1914), together with an impressive battery of other telescopes and cameras, Espin observed 3,800 red stars, thousands of double stars and nebulae, and carried out numerous spectroscopic and photographic studies. On 30 December 1910 he discovered Nova Lacertae. This won him fame that reverberated around the world, and he was awarded the prestigious Jackson–Gwilt Gift and Medal of the Royal Astronomical Society.[42]

In the fullest tradition of the parson–astronomers of the nineteenth century extending into the first third of the twentieth, Espin was independently well off, in addition to which he seemed to have an inexhaustible supply of rich friends and benefactors who made him gifts of instruments and other valuable pieces of equipment. His parish of Tow Law – a rather isolated industrial community in County Durham – also benefited from its eccentric bachelor vicar. His home-made X-ray apparatus was used to examine his injured parishioners, in his garden he built a small sanatorium for consumptives, and he regarded his vicarage, with its vast library and collections of fossils, plants, and instruments, as 'a public house which provided mental stimulants'[43] to the community. He even opened up a gymnasium in the cellar of the vicarage and a rifle range in his garden for use by the Church Lads' Brigade and church choristers. And while Espin – an Oxford MA, Fellow of the Royal Astronomical Society, County Magistrate, ecclesiastical autocrat, and widely-travelled gentleman – never really felt comfortable making house visits to his individual working-class parishioners, his personal generosity towards them as a group cannot be doubted.

It was Espin's slightly younger contemporary, the Revd Theodore Evelyn Reece Phillips, vicar of Headley, Surrey, who was perhaps the last of the clerical astronomical breed. Phillips – who never observed on Saturday nights, no matter how clear, for that was when he wrote his Sunday sermon – was not only the 1918 recipient of the RAS Jackson–Gwilt Gift and Medal and an honorary DSc from his old University of Oxford, but served as the last clerical President of the RAS from 1927 to 1929.[44]

By the time of Espin's death in 1934, however, the clergy of the Church of England were becoming very different from what they had been in the days of Pearson or Webb. Yes, clergymen were still gentlemen of standing, but they had more duties to attend to, and fewer resources to devote to great, privately-pursued intellectual enterprises, while the Church as a corporate body was much harder up financially. The collapse of land values and industrial shares in the teeth of new foreign competition in the years after 1880, the sharp and sustained rise in inflation during the First World War, and the world-wide depression of the 1930s, presented many incumbents with new and more urgent priorities, such as how to heat and staff their once palatial rectories, and how to raise the money necessary to maintain church fabrics. Gone was much of the once-free time of clerical gentlemen. And in the increasingly cash- and resources-starved twentieth-century Church, simply finding leisure for private prayer and spiritual renewal became a problem in its own right, let alone spending time at the eyepiece of a telescope, pondering an ingenious invention, or writing books aimed at encouraging fellow men and women to look towards the sky.

Of course, there are still clerical astronomers. The Vatican Observatory at Castelgandolfo is still staffed by Jesuits,[45] though the increasing secularisation of higher education has largely removed priests from scientific research outside Catholic institutions. However, the founding of bodies such as the predominantly Anglican Society of Ordained Scientists in 1987, and the increasing dialogue between science and religion, indicate that religious belief is by no means absent from the modern astronomical world.

But Thomas William Webb belonged to a different era. Even so, we in the early twenty-first century should be careful not to think of clerical scientists of his generation as becoming priests simply as a way of finding the time and resources to do their science. In the case of Webb, and many more like him, it was a lively Christian faith that motivated his very being, whilst his astronomy became one of several channels – which also included the service of his fellow men and women – through which that faith was expressed.

Notes and references

1 Anthony Trollope, *The Warden* (1855), *Barchester Towers* (1857), *Doctor Thorne* (1858), *Framley Parsonage* (1861), and others. See also Owen Chadwick, *The Victorian Church*, 2 vols. (SCM Press, London, 1966, reprinted 1992) for an excellent and readable social history.

2 Dawes had originally intended to train for the Anglican priesthood, but had his reservations not about Christianity but about the constitutional and governmental basis of the Church of England. He was brought to Congregationalism by Dr Raffles, a minister of Liverpool. See obituary, in *Monthly Notices of the Royal Astronomical Society*, **29** (1869), 116–20.

3 For anti-Catholic feeling, see Chadwick, *The Victorian Church,* Part I, 1829–1859 (n. 1), 1–24. Charles Waterton (1782–1865), the distinguished South American explorer, ornithologist and naturalist, came from an ancient Roman Catholic Recusant family of Yorkshire, and suffered Protestant exclusion simply because he and his ancestors failed to recognise the sixteenth-century Reformation. Men like Waterton were English gentlemen of impeccable standing, yet excluded by their faith from Parliament, magistracies, army careers, and the other professions of their caste. Waterton, however, turned to the unrestricted profession of science. Gilbert Phelps, *Squire Waterton* (EP Publishing, Wakefield, 1976).

4 Temple Chevallier to Dr Corrie, 30 March 1839, for remarks about Ushaw Jesuits: Durham University, Add MSS 837 (No. 38). Also (in the same Durham manuscript sequence) Corrie to Chevallier, 14 May 1839, raging against the insincerity of Catholics and the folly of the 'God-denying Act of 1829' (Catholic Emancipation) (No. 40). For Dr Gilly's explosion fears, see Chevallier to Corrie, 11 March 1836 (No. 12). Also Allan Chapman, *The Victorian Amateur Astronomer. Independent Astronomical Research in Britain, 1820–1920* (Praxis–Wiley, Chichester, 1998), pp. 20–2. For Chevallier in the wider Durham world, see Edgar Jones, *University College Durham: A Social History* (Edgar Jones, Aberystwyth, 1996), pp. 75–7.

5 George Bishop, 'Stephen Perry (1833–1889): Forgotten Jesuit scientist and educator', *Journal of the British Astronomical Association*, **89**, 5 (1979), 473–84.

6 Oxford and Cambridge lost their Anglican exclusiveness in 1871: C.E. Mallet, *A History of Oxford University, III* (Barnes and Noble, New York, and Methuen, London, 1927; reprinted, 1968), p. 332.

7 'Universities', *Encyclopaedia Britannica*, **22** (fourteenth edition, 1930), 862–81; see 873–4, and 'University Colleges', 880–1.
8 Thomas Romney Robinson, 'Report of the Armagh Observatory, 1869', cited in J.A. Bennett, *Church and State in Ireland: 200 Years of the Armagh Observatory* (Armagh Observatory and Institute of Irish Studies, Queen's University, Belfast, 1990), pp. 146–7; see also Chapman (n. 4), p. 26.
9 'Universities' (n. 7), 873–4.
10 T.E.R. Phillips, obituary of Thomas Henry Espinall Compton Espin, *Monthly Notices of the Royal Astronomical Society*, **95**, 4 (February 1935), 319–22. Also *The Star-Gazer of Tow Law* (Tow Law History Group, County Durham, 1992), 8.
11 John Couch Adams' father, Thomas, was a prominent figure in the local Wesleyan movement in Cornwall, organising the chapel and prayer meetings at Tregeare. Yet he was also on very friendly terms with the Anglican vicar, and even served as Church Warden of the parish church. One can see how, with this background, J.C. Adams – who was a devout Christian throughout his life – could move effortlessly between Cornish Wesleyan chapels and Anglican Cambridge. See George Adams, 'Reminiscences of our Family' (1892), printed in H.M. Harrison, *Voyager in Time and Space. The Life of John Couch Adams, Cambridge Astronomer* (Book Guild, Lewes, 1994), pp. 12 and 248. For Adams' obituary by James W.L. Glaisher, which does not discuss his religious beliefs, see *Monthly Notices of the Royal Astronomical Society*, **53**, 4 (February 1893), 184–209.
12 Sir William Huggins, obituary of William Lassell, *Monthly Notices of the Royal Astronomical Society*, **41**, 4 (1881), 188–91. Like several other obituarists, Huggins mentions Lassell's early education at the Duke's Alley Congregationalist Chapel School, Bolton, Lancashire, and at the Rochdale Academy. That Lassell remained a Congregationalist throughout his life is also confirmed by his obituarists specifying his burial, at Maidenhead, by the Congregational Rite in 1881.
13 For St Omer, Douai, Downside and Stonyhurst, see articles in F.L. Cross and E.A. Livingstone, *The Oxford Dictionary of the Christian Church*, revised edition (Oxford University Press, 1997). Charles Waterton (see n. 3, p. 19) was one of the first group of English Roman Catholic boys to return from Liège to the new school at Stonyhurst in 1794.
14 Several modern scholars have examined the work of Father Nicholas Callan (1799–1864), inventor of the 'Maynooth Battery', as well as significant early researches into electromagnetic induction. A good overall account of Father Callan's work was given by Mary Mulvihill in 'An Irishwoman's Diary', *The Irish Times*, 31 August 2000, p. 17 (*The Irish Times* can be consulted on the Internet). I am indebted to Dr Mary Brück of Edinburgh for sending me a copy of the above article.
15 David Gavine, 'Thomas Dick (1774–1857) and the Plurality of Worlds', *Journal of the Astronomical Society of Edinburgh*, **28** (September 1992), 4–10. Also Gavine, 'Astronomy in Scotland, 1745–1900', unpublished Open University PhD thesis (1982), 304-14. Also William Joseph Astore,

Observing God: Thomas Dick, Evangelicalism and Popular Science in Victorian Britain and America (Ashgate, Aldershot, 2002).

16 Samuel Smiles, *Men of Invention and Industry* (London, 1884), p. 329.

17 Smiles (n. 16), pp. 361–9. *The Solar System* (1846), published in Welsh as *Y Dorspath Heulawg* (Liverpool, 1850), and *The Christian Philosopher*, published in Welsh as *Yr Anianydd Christianogol* (1842) were two of Dick's astronomical books that were very influential in Wales. See also Chapman (n. 4), pp. 209–13, for more on Jones, and Dick's books in Wales.

18 The Barchester Deanery was worth £2,000 a year, with a fine house and fifteen acres of ground, though the Ecclesiastical Commissioners were to reduce the stipend to £1,200: Trollope, *Barchester Towers* (1857), Chapter 31. *Crockford's Clerical Directory* (fourth edition, 1868), p. 424, records Webb's parish of Hardwicke as worth only £150 p.a. In *Crockford's* (1883), p. 1220, the living had been augmented to £233. A variety of valuations for the Hardwicke living appear in various local directories: for example, *Littlebury Gazeteer & Directory* (1867), £143; *Kelly's Post Office Directory* (1879), £228. I am indebted to Janet Robinson for the local sources about the Hardwicke finances in Webb's time. What remains, however, is that the incumbent of Hardwicke was poorly paid, and Webb clearly had private resources to live as he did. Webb's social standing and means are also looked at in n. 27. In his Will, signed 12 March 1885 (and subsequently proved in Hereford on 17 June 1885, following his death on 19 May 1885), Thomas W. Webb left £16,986 5s 1d, much of which was expended on bequests to charities, friends and servants.

19 Trollope's *Barchester Towers*, Chapter 9, contains one of the most famous cases of prolonged clerical absenteeism in English literature: that of the Revd Dr Vesey Stanhope, who had left his Cathedral appointment and several parishes to recover from a sore throat in Italy – and had stayed there for twelve years, while still enjoying all of his Barchester revenues. For real-life concern about clerical absenteeism, see Chadwick, *The Victorian Church*, Part I (n. 1), p. 34, where it is recorded that (in spite of some suspected exaggeration in the statistics) of 10,533 Anglican parishes accounted for in 1827, only 4,413 had incumbents who actually resided in their parsonages and ministered to their flocks.

20 Flora Thompson (1876–1947), *Lark Rise to Candleford* (1945), Chapter 14, 'To Church on Sunday', (Penguin Modern Classics edition, 1979), pp. 209–29. Pp. 228–9 relate the rural ministry and sudden death at the altar of the elderly curate the Revd Alfred Augustus Peregrine Marley, probably during the 1890s. I have searched for Mr Marley in Joseph Foster's *Alumni Oxonienses ... 1715–1886* (London, 1888) and J. A. Venn's *Alumni Cantabrigienses ... 1752–1900* (Cambridge University Press, 1951), but can find no reference to his matriculation at either Oxford or Cambridge.

21 Roger Langdon, *The Life of Roger Langdon told by himself, with Additions by his Daughter Ellen* (Elliot Stock, London, 1909), p. 82.

22 T.H.E.C. Espin, 'T.W. Webb – A Reminiscence', in *Celestial Objects for Common Telescopes* (fifth edition, posthumously published, revised by Espin, London, 1893), p. xv.

23 Gladstone – who was probably Britain's most devout Prime Minister – was very much aware of this sense of duty, most clearly expressed in his fifty-year 'social outreach' to 'save' fallen women, in which he was supported by his wife Catherine. Mrs Gladstone was involved with charity work. Roy Jenkins, *Gladstone* (Macmillan, 1995), pp. 104–10; 241 note.

24 'Hester Periam Hawkins', obituary, *Monthly Notices of the Royal Astronomical Society*, **89** (February 1929), 308–9; see also Chapman (n. 4), pp. 287–9 and 402, for further sources.

25 Rita Collins, *The Webb Family* (Salisbury, 2001), a privately-printed volume of biographical, genealogical and photographic sources on the Webbs. I am indebted to Mrs Collins for her generous gift of a copy of this volume, and for other help.

26 Kilvert, *Diary*. For Webb references, see 27 February 1870, March Eve 1870, March Day 1870, 7 July 1870, 29 August 1870, 21 May 1871, 29 May 1871, 11 August 1871, 26 September 1871, 15 January 1877, 27 September 1878 and 21 November l878.

27 For the number of Webb's live-in servants see the 1861 Census return for Hardwicke parsonage. At his death Webb left two carriages, farm vehicles, and at least four fine horses. See Chapter 3 for further information.

28 'The Rev William Pearson, LL.D., F.R.S., F.R.A.S., etc ... ', obituary, *Monthly Notices of the Royal Astronomical Society*, **8** (1848), 69–74. I am indebted to Miss Wendy Atkins, of Sleaford, Lincolnshire, for biographical information on Pearson from local archival sources.

29 J.L.E. Dreyer and H.H. Turner (eds.), *History of the Royal Astronomical Society, 1820–1920* (RAS, 1923, and Blackwell Scientific Publications, 1987), pp. 1–3, 21, 24 and 83.

30 Pearson's thoroughgoing approach to practical astronomy is best displayed in his *Practical Astronomy*, 2 vols. (1829).

31 Ada Pritchard (ed.), *Charles Pritchard, D.D., F.R.S., F.R.A.S., F.R.G.S., Late Savilian Professor of Astronomy in the University of Oxford. Memoir of his Life* (London, 1897): 'Theological Work – by the Lord Bishop of Worcester', pp. 171–212.

32 E.B. K[nobel?], obituary of Stephen Joseph Perry, *Monthly Notices of the Royal Astronomical Society*, **50**, 4 (February 1890), 168–75. Also Bishop (n. 5), 473–84.

33 The obituary of Perry (n. 32), 168, relates that he became Professor of Mathematics and Astronomy at Stonyhurst College in 1860. He completed his theological studies at St Beuno's in North Wales, was ordained Priest in 1866, and returned to take charge of Stonyhurst College Observatory in 1868. The obituary also relates the circumstances of his death.

34 Elizabeth Brown (written anonymously), *In Pursuit of a Shadow, by a Lady Astronomer* (London, 1887); see pp. 89–112 for intended yet clouded-out observational work with Father Perry, Dr Ralph Copeland and Prof Theodor A. Bredichin in Russia.

35 *The History of the Manchester Astronomical Society (The First Hundred Years)* (UMIST, Manchester, 1992), pp, 1–46, makes several references to Jesuit officers of the Society.

36 Deborah Jean Warner, *Alvan Clark & Sons. Artists in Optics* (Smithsonian, Washington, 1968; new edition, 1995), 107.
37 RAS MS Webb 5, 1 May 1869.
38 Henry Cooper Key, 'On a mode of Figuring Glass Specula for the Newtonian Telescope', *Monthly Notices of the Royal Astronomical Society*, **23** (1863), 199–202.
39 R.A. Marriott, 'The Life and Legacy of G. H. With, 1827–1904', *Journal of the British Astronomical Association*, **106**, 5 (1996), 257–64.
40 Edward Lyon Berthon, 'On Observatories', *English Mechanic*, 342 (13 October 1871), 83–4.
41 *Nelson's Encyclopaedia*, **19** (London, *c*.1911), under 'Romsey', mentions that in 1911 'The construction of Berthon collapsible boats is a special industry' within the town of 4,671 souls. Also E.L. Berthon, *A Retrospect of Eight Decades* (Bell, London, 1899) for a photograph of a 28-foot Berthon boat, being rowed by 75 men. The Berthon Memorial Window in Romsey Abbey includes as insets a stained-glass portrait of Berthon, a boat, a telescope, and astronomical bodies within a greater composition based upon a setting of the young Christ in the carpenter's shop of Joseph with the Virgin Mary, and surrounded by various tools. It is reproduced in Judy Walker's guidebook *Romsey Abbey* (English Life Publications, Derby, 1988), p. 16.
42 Revd T.E. Espin, 'Note on the Visual Spectrum of the uncatalogued Red Star in Lacerta', *Monthly Notices of the Royal Astronomical Society*, **71**, 3 (January 1911), 189–200. Also, President of the RAS to Espin on the presentation of the Jackson–Gwilt Medal and Gift, *Monthly Notices of the Royal Astronomical Society*, **73**, 4 (February 1913), 330.
43 H.W. Davis, obituary of T.H.E.C. Espin, *Journal of the British Astronomical Association*, **45**, 3 (January 1935), 128. See also other Espin biographical and obituary publications listed in n. 10.
44 President of the RAS to the Revd T.E.R. Phillips, on the presentation of the Jackson–Gwilt Medal and Gift, *Monthly Notices of the Royal Astronomical Society*, **78**, 4 (February 1918), 337–9. W.H. Steavenson, obituary of T.E.R. Phillips, *Monthly Notices of the Royal Astronomical Society*, **103**, 2 (1943), 70–2.
45 Sabino Maffeo, SJ, *In the Service of Nine Popes: 100 Years of the Vatican Observatory*, translated by George V. Coyne, SJ (Pontifical Academy and Specolo Vaticana, 1991). This is an excellent and detailed history of the Jesuit-operated Papal Observatory. Although dealing with the earlier work of Fathers de Vico, Secchi, and others, it concentrates on the Observatory's work in astrophysics in the Vatican City, 1888–1933, and then under the clearer skies of Castelgandolfo, some fourteen miles south of Rome, after 1933. (At Castelgandolfo, the Vatican Observatory is incorporated into the Pope's Summer Residence, which suggests an especially intimate relationship between the Church and astrophysics; but the deterioration of the optical clarity of even the Italian climate has, alas, occasioned the relocation of the Vatican Observatory's most recent research equipment under the desert skies of Tucsan, Arizona. I am indebted for this information to Father George V. Coyne, SJ, Director of

the Vatican Observatory, who kindly hosted my visit to Castelgandolfo in June 2001. The Castelgandolfo Observatory nonetheless actively hosts international conferences in astrophysics, though from its terrace, facing northwards, one can see the thick blue smog of modern pollution that hangs over Rome.)

Chapter 8

Webb's telescopes

Robert A. Marriott

Why my telescope is a superior one by Herschel's own tests!!!

During his lifetime Webb saw dramatic changes in the size, design and quality of telescopes. In his youth he was familar with modest Newtonian reflectors and very small refractors, and eighteenth-century Gregorian telescopes were still being used. There were in existence, of course, a few very large reflectors, but Webb could only have dreamed of owning and using such instruments.

During the 1790s William Herschel had constructed his largest instrument – with a 48-inch mirror of 40-foot focus – and by 1845 Lord Rosse had completed his 72-inch reflector – the Leviathan of Parsonstown. But at that time telescope mirrors were not made of glass. From Isaac Newton's first reflecting telescope of 1668 until the mid-nineteenth century they were made of speculum metal – an alloy of copper and tin. Speculum metal was the best material available, but at its most efficient it had only 70 per cent reflectivity, it tarnished after a relatively short time, the arsenic that was added to increase reflectivity further reduced the working life of the surface, frequent repolishing was necessary, and it was very heavy.

In the latter half of the eighteenth century the design and production of lenses was revolutionised by the invention of the achromatic lens and the introduction of new processes of glass manufacture, and the refractor was beginning to compete with the reflector in capability and performance.

Mirrors

Much is known about the work of amateurs such as John Herschel, James Nasmyth, William Lassell and Lord Rosse, who were sufficiently wealthy to be able to construct and use very large instruments for private research; but history records very little about the activities and more modest efforts of the less affluent amateur instrument-makers of the first half of the nineteenth century. Webb, however, in his diary, recorded his attempts at mirror- and lens-making during the period 1826–31. Many of the notes are in abbreviated form and are often enigmatic, but they are sufficient to reveal his activities and progress, and provide a brief but valuable glimpse into the almost unknown world of the more humble but equally enthusiastic optical practitioners of that period.

Webb's first recorded astronomical observation (of a meteor) was made on 5 January 1818,[1] when he was eleven years old, and two weeks later he observed the Moon. It is evident that he had access to small telescopes belonging to friends, and in 1823 he recorded observations made with a 1.3-inch refractor by Bate (which may have belonged to him). Within a few years, however, he had begun to make his own mirrors and lenses. On 11 April 1826 he 'Got at Chambers [cyclopaedia], and there all manner of information about lenses &c &c &c as to focal lengths, apertures, powers, refraction of turpentine, &c.'; and on 1 May he declared his intent to begin a regular account of observations.

During the late 1820s Webb made several small mirrors up to 6 inches in diameter, although the majority were of 3 inches. The standard materials for making telescope mirrors are carborundum (tungsten carbide) for grinding, and pitch tools and rouge for fine optical polishing. Webb, however, used a variety of materials for the various processes: grindstone, pewter turnings, grit, emery, ivory, pitch, lead, putty, crocus martis (colcothar, or rouge – a brownish red peroxide of iron obtained from iron sulphate) – or whatever was available. He also frequently visited local tradesmen in his efforts to accrue advice and knowledge, and to acquire materials with which to construct a telescope mount and stand.

Eventually he completed the construction of his own Newtonian reflector, and on 9 September 1827 wrote in his diary, in large black letters:

> First Trial ... of Newtonian with fixed specula & eye-piece – And extremely satisfactory, immensely improved. seemed to bear a power of 60 or 70 but the lower screw of great spec. w'd not hold – so c'd not adjust it properly.

He afterwards continued to improve the instrument – and he appears to have been successful. On 23 August 1828 he reported excitedly: 'Why my telescope is a superior one by Herschel's own tests!!!'

Lenses

At that time, refractors were still comparatively small (except for a very few notable exceptions), and lenses were expensive, due to the difficulty of producing large pieces of homogeneous glass with a paucity of bubbles and without striae or other defects. Webb could therefore not afford such an instrument.

A lens is also subject to several abberrations – more than is suffered by a mirror. During the seventeenth century, Isaac Newton attempted to correct aberrations by introducing water between lenses of the same type of glass, and David Gregory suggested that materials of different refractive index might be effective. During the 1730s, Chester Moor Hall commissioned opticians to make him a two-element object-glass. John Dollond repeated and improved these experiments, and afterwards designed two-element objectives with crown and flint glass components, as devised independently by Hall. In 1758 Dollond patented this invention: the achromatic lens.

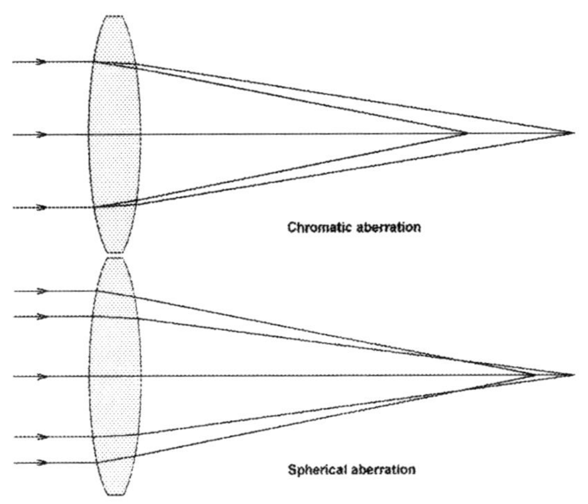

Chromatic and spherical aberration.

In 1787 Dr Robert Blair (*d.*1828), Professor of Practical Astronomy in the University of Edinburgh, began investigations into finding a substitute for flint glass – which was difficult to produce, and more expensive than crown glass – by using various oils and metal salts, in some cases enclosed in separate cells, and in others mixed in one cell between two convex crown glasses. After several years he succeeded in eradicating the secondary spectrum – the very small residue of colour produced by an otherwise 'achromatic' lens – and was also successful in removing spherical aberration. (A lens or mirror subject to spherical aberration is incapable of bringing rays to focus in the same plane normal to the optical axis if the distances of these rays from the axis are different.) Blair consequently applied the term 'aplanatic' ('free from aberration') to his lenses.[2] Several prominent opticians maintained that lenses of this type were not reliable, due to loss of transparency of the fluid by evaporation or crystallisation, or the corrosive action of acids on glass; but Blair disagreed. However, he discontinued this work, and his later attempts to manufacture fluid lenses on a commercial basis proved unsuccessful.

During the 1820s, Peter Barlow (1776–1862), Professor of Mathematics at the Royal Military Academy, Woolwich, also began research and experiments on the use of fluid lenses.[3] In Barlow's design the fluid lens replaced the flint-glass component of a doublet, but was placed well away from the single-lens object-glass along the optical axis. It cost far less than flint glass (which was still difficult to produce in large pieces), and because of its position in the optical train it could also be much smaller. Between its two components it had a gap, into which was introduced a fluid with a refractive index appropriate for correcting the aberrations produced by the object-glass. Barlow determined that the best fluid for this purpose was carbon disulphide, due to its perfect transparency, its absence of colour, and its high refractive index – twice that of flint glass. In 1827 Barlow made a 6-inch system of this type, and in 1829 another of 7.8 inches, with satisfactory results. However, in 1833 his attempt, in cooperation with George Dollond, to produce an 8-inch (at a cost of £157) for the Royal Society proved unsuccessful, and he soon afterwards discontinued his work on fluid lenses.

To form the fluid into a concave lens it was enclosed between two discs of glass, each with the requisite curve but with parallel faces so as to have no refractive or dispersive action. These were applied to the two opposite faces of a third disc, with corresponding curves and with its centre bored out to produce a ring. The three discs and the fluid were then all gently warmed to a temperature higher than any likely to be reached under normal conditions, the ring was placed upon one of the discs, and the other disc was slipped on to one side so as to leave open a small portion of the interior of the ring. The fluid was then poured in, and the upper disc slipped into place. Tin-foil or paper was then cemented around the edges to complete the process. When the fluid cooled its contraction produced a vacuum-bubble, which was kept out of sight by allowing the ring an extra amount of aperture.

By 1826 Webb had begun experiments with various fluids in an attempt to make his own lenses. In April of that year he had learned about the refractive properties of turpentine, and was soon experimenting with its use in an object-glass: 'Focal distance very short, but apparently achromatic.' He may have continued these experiments at the time, but the next notes appear three years later, when on 31 July 1829 he

> ... tried to make a chromatic [*sic*] lens w. spt. turpentine between two eyeglasses – it had not refractive power enough, yet certainly had a good effect. Calculating for Achromatics in which Crown glasses sh'd make the Concave, Water or Alcohol the Convex. C'd not succeed ... Various plans of telescopes.

Throughout this time he continued with his mirror-making, and over the ensuing eighteen months occasionally returned to his fluid-lens experiments. The fluids he tried included aniseed oil, which 'ran at full speed' and 'made the house stink prodigously'; Canada balsam, 'found its refractive power very high'; and turpentine, 'colour well corrected ... Very nearly broke the glasses out of the laundry window'. The only cement referred to was a mixture of gum and pipeclay. During December 1829 he 'tried Bates's [*sic*] lens with Canada balsam ... [and] again w. turpentine';[4] that is, he used the balsam and then the turpentine in a fluid lens to correct the aberrations produced by the single-element lens.

Eventually, in August 1830, he used the chemical which Barlow had determined was best for the purpose:

> Meant to fill the glasses before dinner, but could not have the fire. Looking to the Sulph. C. [sulphuret of carbon – carbon disulphide], perceived it was stopped down w. some queer cement. Took it to Fouracres [a Gloucester chemist]. He said it was a scandalous stopper & had wasted so that only 6 dr[ammes] remained. He gave me another bottle.

As well as its transparency, absence of colour, and high refractive index, carbon disulphide has other characteristics which clarify some of Webb's cryptic notes. It boils at 35° C (lower than body temperature), the vapour can settle for some time or can 'roll' across a surface as if adhering to it, and it smells of rotten cabbage.

One other problem remained, however. With this type of optical system, spherical aberration increases with increased distance between the primary lens and the corrector. At the end of the previous March, while in Oxford, Webb had visited Stephen Rigaud (Professor of Astronomy, and Director of the Radcliffe Observatory) and 'asked him about spherical aberrations, and he most kindly promised to look into it'. Rigaud subsequently provided him with information, but although Webb spent much time in investigating this problem, he did not succeed in removing spherical aberration from this telescope.

It is possible, therefore, that this instrument consisted of a 4-inch single-lens object-glass made by Bate and a carbon disulphide lens made by Webb, who also made the tube, finder and mount, no doubt in consultation with his associates and tradesmen. Writing thirty-five years later, Webb considered its performance to be acceptable, although he equated it with Barlow's attempted 8-inch, which 'proved a failure; and such might be considered my own humble imitation ... It served me, however, for four years, with tolerable achromaticity, but much uncorrected spherical error.'[5]

Only a few fluid-lens refractors were made, and with the production of higher-quality optical glass they effectively became redundant. More than three decades later, however, in 1865, Webb wrote that 'several limpid fluids have since been discovered whose properties might merit investigation, especially chloroform, which from its density seems to promise well'[5] – an indication that even at that time the fluid lens was still considered a viable option.

Dialytes

In the late 1820s, at about the time that Barlow began his experiments, A. Rogers[6] proposed a new design for a dialytic[7] telescope. As with a fluid-lens refractor, a dialyte is one in which achromatism is affected by the positioning of a flint or crown/flint (instead of fluid) correcting lens along the optical axis, at some distance from the crown or plate object-glass. The object-glass is also thinner, and has shallower curves than the crown component of a doublet of equal aperture; and as the more expensive flint component is small there is a large saving in cost, especially with larger instruments. The distance between the object-glass and the corrector is arbitrary, and a smaller lens placed nearer the focus serves the same purpose; but the disadvantage (as with a fluid lens) is that spherical aberration increases with greater distance. Rogers' design enabled a 3-inch flint/crown lens to correct the colour produced by a 9-inch crown glass of 14 feet focus.

This design appealed to Webb, and in 1832 he asked his father to consult with George Dollond about the possibility of making such an instrument. Dollond replied 'very sensibly and to the point', Webb later told his father, 'but we shall never "make a deal" of it. He will not come to my terms nor I to his, to the tune, I daresay, of £50 or £60'.[8] Dollond did not consider that such an optical system would have any superiority over the usual achromatic object-glasses, nor would it lessen the cost, and the planned instrument was not produced. Webb also considered making 'an attempt with Gilbert or Bate', but then

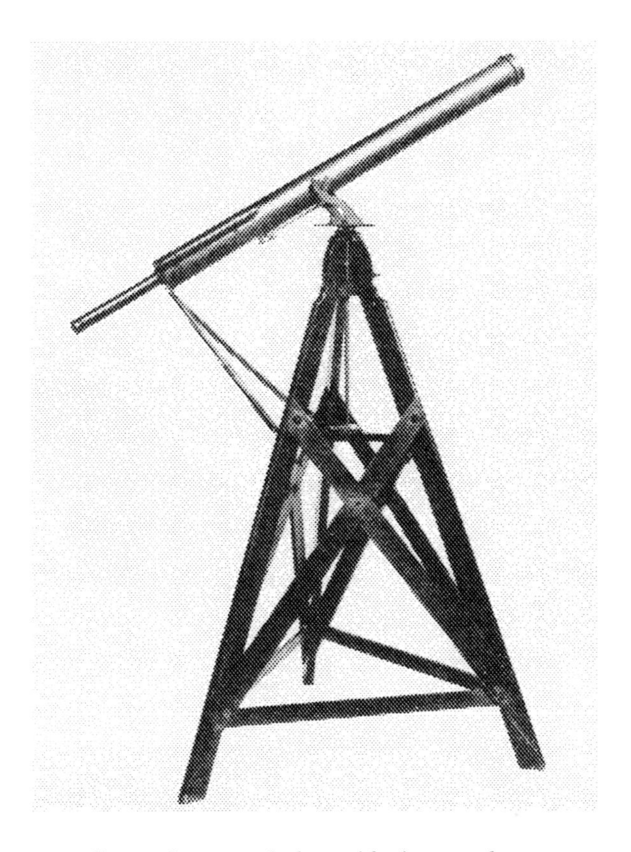

A Tulley refractor of the mid-nineteenth century (see p. 129).

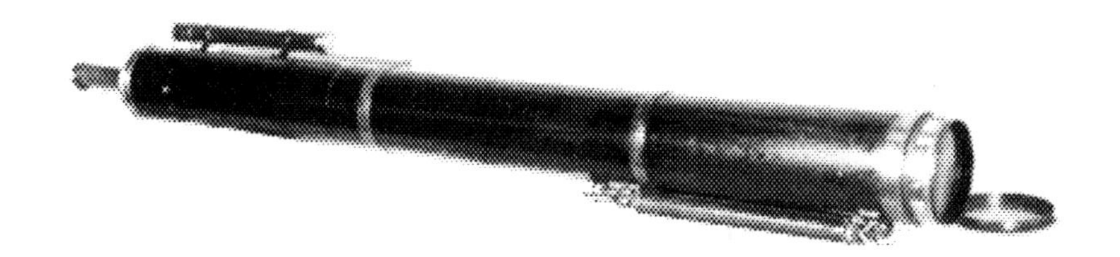

The $3\frac{1}{2}$-inch refractor by Dollond, loaned to Webb by the Royal Astronomical Society (see p. 130).

The $9\frac{1}{3}$-inch With–Berthon reflector belonging to the British Astronomical Association.

A 'Romsey' observatory, designed by Revd E.L. Berthon (see p. 139).

decided against it, as 'they know no more of the requisite curves than myself, and we should all be in the wood together ... The wisest course is to discount such schemes and expectations altogether'.

In Great Britain the dialyte never attained popularity, and only a few were produced – even though John Herschel considered it 'a beautiful invention, highly deserving further trial'. The best of them were produced by G.S. Plössl in Vienna during the 1840s and 1850s; and of these, the largest was an 11-inch with a focus of 11¾ feet. Dialytes are consequently very rare, although there now exists a notable example. In 1865, Webb wrote that 'in theory, an object-lens of plate glass, however large, may have its colour corrected by a disc of flint glass, however small';[9] and 135 years later the English amateur astronomer John Wall completed the construction of a dialyte of 30 inches aperture.[10]

3.7-inch refractor by Tulley

Webb's last notes on mirror-making are dated September 1831, about a year after he completed his fluid-lens refractor, which he used for four years. Then, on 11 June 1834 he recorded his excitement with the 'Prospect of a capital telescope!' – a 3.7-inch refractor by 'the younger Tulley' (William or Thomas), purchased for him by his father. On 3 July it duly arrived, 'and a splendid thing indeed it was. The good Lord give me the grace to make a Christian use of it.' At first he accustomed himself to its use, and on 22 July he subjected it to

> ... a thorough trial, to do which I sat up, & knelt down on the gravel path, till past 1. The result was most completely & unexpectedly satisfactory ... for very little did I think I had got a first-rate instrument when I received it.

A decade later, in 1844, he purchased a copy of W.H. Smyth's newly published two-volume work, *A Cycle of Celestial Objects* – probably the most eloquently written book on astronomy ever produced. The second volume of this work is the Bedford Catalogue – the result of Smyth's survey of the entire sky visible from his home in Bedford, carried out with a 5.9-inch Tulley refractor over a period of more than a decade. This was not a 'common telescope' (it was one of the largest refractors of its day), and Webb therefore decided to carry out a similar survey, more suitable for smaller instruments. His 3.7-inch refractor was his primary instrument for twenty-four years, and was the 'common telescope' which he used to compile his *Celestial Objects for*

Common Telescopes (first published in 1859). Webb dedicated this book to Smyth, and at the same time pointed out that the *Cycle*, by 'its very superiority, to say nothing of its bulk and cost [£5], renders it more suitable for [the advanced amateur's] purpose, than for the humble beginnings which are now in view'.

3½-inch refractor by Dollond

Apart from the instruments which he owned, and the occasional small telescope briefly loaned to him during his youth, Webb also borrowed at least one other telescope: a Dollond refractor. Throughout the nineteenth century until the 1920s, instruments in the collection of the Royal Astronomical Society were placed on loan to Fellows. One of these instruments was a triple achromat by Peter Dollond, who produced the first of these lenses around 1763. They were much more difficult to make than doublets, however, and it is probable that only about a dozen or so were made over a 20–30-year period. Even as early as 1818, William Kitchiner wrote that 'these instruments are now extremely rare to be met with, as they are much more difficult to make than double object-glasses'.[11]

In 1771 Dollond supplied a 3½-inch f/12.7 triple achromatic refractor to Francis Wollaston, and in 1828 Wollaston's son, William Hyde Wollaston,[12] presented this instrument to the Astronomical Society of London (later the Royal Astronomical Society), shortly before his death. Wollaston had expressed his hope that the Society would 'not keep it useless, but lend it, or give it if they think proper, to any industrious member'.[13] His hope was realised to a certain extent, and the telescope was loaned to Thomas Maclear, of Biggleswade, from 1829 to 1833, and to H.C. Schumacher, of the Vienna Observatory, from 1839 to the early 1850s. Six years later – and twenty-five years after Webb had first received and read some of Schumacher's writings – it was at Hardwicke.

On 27 April 1852 Webb wrote to the Secretary of the Royal Astronomical Society, in a formal and almost ingratiating style:

> I have observed in the Annual Report that there are at the present time some instruments in the apartments of the Society, which, if my memory does not mislead me as to the Rules, the Fellows are permitted to borrow at the pleasure of the Council. I do not know whether there may be anything unusual or informal in such a request from a Fellow so recently elected, and not yet formally admitted to the Society – but if I may hope that such an application would not be thought to involve any irregularity or impropriety, I would beg permission to state that the Council would

> confer a great obligation upon me by entrusting one of the telescopes to me during their pleasure. I have for nearly eighteen years been in the habit of using a five-feet Achromatic by Tulley, and I have myself ground and polished several specula – circumstances to which I take the liberty of referring, simply as perhaps serving to shew that due care would be taken of the instrument, should the Council do me the favour to entrust it to my hands.[14]

Webb was supplied with information on available instruments in the collection, but there was no immediate action. Three years later, on 7 May 1855, he submitted a gentle reminder concerning his request, 'which I believe was favourably received', and decided that Wollaston's 3½-inch Dollond 'alone seemed likely to be of use to me'. (This is perhaps a little surprising, as it was smaller than his Tulley refractor.)

The following December, Webb acknowledged receipt of the telescope. On 5 January 1857 he told the Secretary of the RAS: 'I do not think very highly of the Wollaston Telescope ... I cannot pronounce it first-rate.' Later, however, he declared it to be 'very fine against my Tulley; & though there is no *very wide* discrepancy, I think mine beats it'.[15] He made little use of it, however, and in 1859 he returned it to the RAS. In later years it was loaned to G.B. Airy (the Astronomer Royal) for a transit of Venus expedition in either 1874 or 1882 (the record is not specific), and finally to the meteorologist Richard Inwards, from 1889 to 1901.

In 1956 the Royal Astronomical Society presented the instrument to the British Astronomical Association, and in 1980 Horace E. Dall subjected it to a thorough examination and optical testing. Significantly, his tests revealed several small errors, and tests on the concave surfaces of the flint lens showed 'a moderate amount of astigmatism, believed due to slight bending or warpage of the whole lens – possibly from mounting stress'.[16]

Since 1982 this instrument has resided in the collection of the Science Museum, South Kensington.

5½-inch object-glass by Clark

After selling his Tulley refractor in January 1858, Webb was without a sizeable telescope, although during this period some of his observations were made with 'Mrs Webb's little Bardou, 27.8 in focus & 2.2 aperture, power about 43'. A few months later, however, he took delivery of a 5½-inch aperture 7-foot focus object-glass.

By the late 1840s the refracting telescope was again beginning to gain ground. Optical companies such as Merz, Repsold, Steinheil, Dollond, Cooke and Wray were producing larger instruments of much higher quality, and other new optical companies were soon to enter the market. In 1847 a 15-inch Merz refractor was installed at Harvard College Observatory, but there was concern in America that large instruments had to be imported from Europe. One of those discontent with this state of affairs was a portrait painter in Cambridge, Massachussetts: Alvan Clark.

In the early 1840s Clark had begun a collaboration with his sons, George Bassett Clark and Alvan Graham Clark, to produce telescope mirrors and object-glasses. In 1851 he wrote to William Rutter Dawes (at Wateringbury, near Maidstone) about the close double stars he had observed with his own 4¾-inch and 5¼-inch lenses. Dawes subseqently purchased five object-glasses from Clark, the largest of which was an 8¼-inch, purchased for £200 in 1859.[17] After Clark's visit to England in 1859, and because of Dawes' recommendations, the reputation of Alvan Clark & Sons was secure, and the company went on to produce refractors which are still the largest in the world.

Dawes' recommendations also reached Hardwicke. On 3 February 1858, Webb wrote to the Secretary of the Royal Astronomical Society:

> I have sold my Achromatic, and ordered one of Alvan Clark's Object Glasses. He tells me I am to have the choice of two, of 5 and 5½ inches in diameter – and they will probably be ready by May.

By June, however, he had not received it, although it was 'approaching completion ... and promising great excellence.' But it duly arrived. It cost Webb £47 5s 2d (plus 17s carriage from Boston to Hardwicke) – and he then had to acquire or manufacture a tube and mount for it. On 27 September he told Arthur C. Ranyard:

> I should have been glad to put it to work before this time, but the people whom I employed in Birmingham [to make] a tube for it – 7 foot long – did it so badly that I had to send it back, and have not yet received it again.

Three days later he recorded in his journal:

> First Trial of the Great Object Glass by Alvan Clark, 5½ inches clear aperture; fitted up temporarily in an old square deal tube. Its performance, in the utter absence of centring, appeared to be admirable.[18]

On this occasion he observed Donati's comet, which, after its discovery on 2 June 1858, appeared as a bright naked-eye object in September and October of that year.

> Having lately received an object-glass from Mr Alvan Clark, of 5½ inches clear aperture, of which he speaks highly, but which I have as yet had no opportunity of testing, I was anxious to turn it upon the recent magnificent comet; but a disappointment relative to the tube obliged me to content myself with a very rough temporary mounting, so that my observations were made under altogether unfavourable circumstances as to convenience; their results are, however, I believe, worthy of confidence.

These observations, carried out over several nights, were indeed considered worthy, and were later published in conjunction with the notes and drawings of W.R. Dawes, William Lassell, James Challis, James Breen, Warren De La Rue, and several other notable observers.[19]

By the following year Webb had still not solved the problem of providing an adequate mount for this instrument. On observing the Moon on 7 February 1859 the telescope was 'out of adjustment' (although two weeks later he observed Saturn with it), and five months later, on 12 July, the night was

> ... fine enough for a trial of Alvan Clark's Object Glass on the stars; but stand miserably unsteady – tied up with bits of string. I used an old negative eyepiece purchased of Miss Tulley, power about 440.

And in a letter to Ranyard a month later (12 August) he wrote of his want of a steady stand as well as of time, 'which is a very scarce commodity with me'.[20] It seems that Webb never overcame the difficulty of providing a satisfactory mount for the Clark object-glass, and there is no record of his using it after 1865. In 1886, shortly after his death, it passed to Stonyhurst College Observatory, at Clitheroe, Lancashire.

By 1863, however, Webb had taken advantage of new technology.

The advent of silver-on-glass mirrors

Throughout his adventures in mirror-making, Webb worked with speculum metal. In principle there was no reason why glass should not have been used for smaller mirrors; but the surface of the glass requires an extremely thin coating of highly reflective material precisely matching the optically worked surface, and the technology to exact this process was not available. Experiments in depositing metal films on speculum-metal mirrors and glass had been attempted as early as the mid-eighteenth century, but for the following hundred years or so the process was used only for domestic or aesthetic purposes. Then – almost suddenly – new technology became available.

In 1855 the German chemist Justus von Liebig succeeded in refining the process of producing metallic surfaces fine enough for optical purposes – and yet he did not apply his process to telescope mirrors. However, the following year this was successfully and independently accomplished by C.A. von Steinheil in Munich and J.B.L. Foucault in Paris. Moreover, these efforts were not inadventurous, as Foucault soon produced a 20-inch silver-on-glass mirror.

Thus it was no longer necessary to acquire pure copper, tin and arsenic, or to undergo the risk of using a furnace, crucibles, moulds and associated equipment. Instead it became possible to obtain glass blanks and grinding and polishing materials, while the silver coating – with a reflectivity exceeding 90 per cent (far greater than speculum metal, but less than aluminium, which is now used) – could be applied by the maker or by a qualified chemist. A silver coating lasted much longer than a speculum-metal surface, but when the coating deteriorated, the surface of the glass could be chemically cleaned and recoated without interfering with the optical figure – and a glass mirror could be more easily handled than a much heavier speculum-metal mirror of equal size.

After Steinheil's and Foucault's success, the first silver-on-glass mirrors were made by the Revd Henry Cooper Key, vicar of Stretton Sugwas, near Hereford. Key began in 1859, and succeeded in first producing a 12-inch f/10 mirror (later sold to David Gill[21] in Aberdeen), and afterwards, an 18¼-inch of 11-foot focus (with a mount, weighing nearly two tons, designed by the Revd E.L. Berthon[22]). In 1863 Key published an account of his methods, incorporating his newly designed graduated tools and the use of a machine similar to that of Lord Rosse.[23]

Key's work came to the notice of George Henry With, who in 1851 had, at the age of 24, been appointed Master (headmaster) of the Blue Coat Schools (one for boys and one for girls) in Hereford. With was one of a new breed – a teacher who was qualified for, and capable of, teaching – and he was keenly interested in science in several fields.[24] He was obviously influenced by Key, and towards the end of 1862 he began work on his own mirrors. In March 1863 Key reported that With had

> ... in his small intervals of leisure time during the last three months ... completed three or four glass specula of the highest class, 6¼-inches aperture and 6-feet focus, and 5½-inches aperture and 4½-feet focus ... One of

George Henry With, 1827–1904.

Sir William Huggins with the 15-inch Grubb refractor loaned to him by the Royal Society (see p. 141).

> his 5½-inches specula ... is presumed to be perfect: not the least trace of error of figure can be detected over the entire surface after the strictest scrutiny.[25]

The following August, Webb told the Secretary of the Royal Astronomical Society that With had made

> ... but few specula as yet, but his work is very good, as I can testify. I have had a 5¼-inch [5½-inch] silvered Newtonian on trial lately, and it does its duty well ... I think it must be fairly equal to a 4-inch achromatic or more – & he will yet, I am persuaded, do better still.[26]

By this time With had already begun to sell his mirrors – one of which had been acquired by the Revd Matthews, of Great Yarmouth. Nothing is known of Matthews, except that he challenged the young George Calver to make a mirror of equal quality. Calver rose to the challenge, became England's third producer of silver-on-glass mirrors, and soon began to manufacture instruments of all types, from simple altazimuth mounts to equatorial mounts with setting circles and clock drives.

In 1863 Webb borrowed one of With's 5½-inch mirrors to try out, and the following year he purchased an 8-inch mirror from him. (Webb's journals contain only occasional references to this instrument, although he apparently used it in conjunction with the $5\frac{1}{2}$-inch Clark object-glass to compare their performance.) He continued to praise the maker who, he said, was

> ... going on admirably with his specula. Those of $6\frac{1}{4}$ inches and $5\frac{1}{2}$-foot [6-foot] focus have, I hear, most marvellous definition. He has just sold a splendid thing of $10\frac{1}{8}$ inches, the performance of which greatly delighted me.[27]

In November 1863 Webb advised Arthur Ranyard on the purchase of a telescope, and provided him with much useful information on the various types, together with details of manufacturers and prices:

> It might be worth your while to consider, before finally deciding, the comparative merits of the silvered glass reflector. You have probably heard of this beautiful instrument ... At present it is only in the hands of amateur makers, but their success has been remarkable. One of at least 8 inches clear aperture may be purchased in Hereford for about £26 or £27. As far as *looks* go, it is certainly a very common and clumsy-looking affair – being merely a great square tube of stained deal, mounted on a very plain wooden stand – and if you regard appearances I could not say much for it. But the Newtonian reflector, under any circumstances, is a singular looking instrument.[28]

The influence of these pioneers of the new method also extended to those not usually associated with astronomical activities. Besides being a first-rate optician With was also a very able chemist, specialising in soil science and in particular the production and applications of fertilisers for agriculture and horticulture. He was also a Fellow of the Chemical Society (later the Royal Society of Chemistry) and a member of Trinity College, Dublin, and counted Thomas Huxley, Michael Faraday, John Tyndall and Edward Frankland among his associates.

Key would no doubt also have acquired skills in chemistry to be able to silver his mirrors, and in 1863 he advised Edward Frankland on methods of mirror-making. He afterwards wrote:

> Dr Frankland immediately and readily figured a speculum of 7 inches aperture ... He, however, expects superior results from a machine-made speculum, and speaks in the highest terms of my suggestion of the graduated pitch-tool.[29]

And Frankland, being a chemist, would not have encountered any difficulty with the silvering process. Indeed, in his advice to Ranyard, Webb pointed out that

> ... the film of silver is also liable to tarnish, unless it is taken care of; but with due precaution Dr Frankland of the Royal Institution, who has made one of these instruments, finds this will not happen – but if it does, resilvering may always be effected, without the least injury to the figure of the glass.[30]

Key advised that mirrors should be silvered

> ... by Liebig's process; and the coat should be rather thin, as, if thick, it is liable to lose its fine figure by rubbing up. The silver will never tarnish if kept dry. A small bag should be kept in the same box with it, containing sawdust which has been saturated with chloride of calcium, and dried.[31]

Silver-on-glass mirrors quickly became very popular. They could be purchased and installed in a home-made tube and mount, according to taste, skill, available materials and facilities, and pecuniary means; and very soon, complete instruments with silver-on-glass optics could be purchased. Furthermore, such an instrument cost far less than a refractor of the same aperture. As Webb later stated: 'A trifling outlay will often procure them of excellent quality, at second-hand; and many are only waiting to be called into action'[32] – the 'trifling outlay', of course, being dependent on individual circumstances. With was selling his mirrors, Calver had set up business – and then John Browning entered the arena.

Browning's family had specialised in the manufacture of nautical instruments since 1760, but on succeeding to the business in the early 1860s he decided to concentrate on the production of purely scientific instruments – especially telescopes and spectroscopic equipment. By the late 1860s he was employing more than seventy workers at his factory in central London, and besides producing complete instruments with his own optics he also supplied telescopes with optics by With. In 1867 he published *A Plea for Reflectors* (later republished several times), extolling their virtues and advantages, and recommending With's mirrors.

The market rapidly expanded, and instruments were soon being sold in Britain and around the world. With's largest mirror was an 18-inch, dated 1877 (although Webb noted that it was 'on the tool' in April 1875[33]). He usually signed and dated his mirrors, frequently added an inscription in Latin, Greek or Hebrew – sometimes all three – and often labelled them 'absolute perfection' or 'wonderful perfection', although he was the first to admit that not all of them were of the same high standard. Those he considered perfect were kept as 'choice reserves', to be sold many years later.

Of these first exponents of silver-on-glass, Calver and Browning produced instruments *en masse*, while With made only mirrors during the time that he could afford. Key, however, took a different approach, and it appears that he made very few (perhaps only two) mirrors – or, at least, no records have survived – although he used his 18¼-inch for many years. Nothing is known of his later life, and he died, at the age of 60, in 1879. Webb was obviously affected by the loss of one who was not only a close friend, but also a fellow astronomer and cleric. In March 1880 he urged that notice of Key's instruments should be posted at the Royal Astronomical Society, and told Ranyard:

> I miss poor Key very much. His noble 18-inch [*sic*] reflector & Obs[ervator]y will soon be on the market, & going I fear too cheap. If you sh'd know of a likely purchaser please bear it in mind.[34]

The 18¼-inch soon found a new home with the Revd Jevon J. Muschamp Perry, vicar of St Paul's, Alnwick, Northumberland. Perry already owned a 6½-inch reflector by Calver, and shortly afterwards he also purchased With's 18-inch mirror, which he considered to be 'a magnificent specimen of that eminent maker's handiwork'. His enthusiasm, however, was short-lived. In 1882 he sold the 18-inch to Nathaniel E. Green (1823–1899) – a highly skilled planetary and lunar observer and a professional artist.[35] Green already had three other With mirrors, and in 1897 he presented the 18-inch to the British Astronomical Association – and it is still in use. Perry remains an obscure figure in the annals of astronomy.

Over a period of around twenty-five years With produced about two hundred mirrors. In 1887 he decided to sell his entire stock of 'reserves' of more than a hundred of them, and therefore issued a four-page broadsheet entitled *Sale of Mr With's Choicest Reserves of Silvered Glass Specula for Newtonian Telescopes*, with the heading 'A Remarkable Opportunity for Amateur Astronomers'. Five sizes were offered, each over a range of focal ratios, for as little as a fifth of the original price. There is no evidence indicating the success of this offer, but it is probable that With's reputation and the comparative paucity of his mirrors would have provoked an immediate response. With the sale of his 'reserves' his mirror-making activities probably ceased, although he continued his work in education and as a lay reader in Hereford Cathedral.

Webb was almost of the same generation as Browning, who died, aged 90, in 1905; but he was a generation older than With – to whom he had offered so much encouragement – and Calver, who indirectly fell under his influence, and who, throughout a long career, maintained a business with several employees and reputedly produced several thousand instruments. With died at the age of 77 in 1904, and Calver died, aged 93, in 1927.[36] Moreover, Webb's young friend T.H.E.C. Espin[37] acquired two large Calver reflectors – a $17\frac{1}{4}$-inch and a 24-inch (which are still in use) – and With and Espin were two of Webb's executors.

Over the decades, some of With's mirrors have disappeared, some have been used intermittently or constantly, and others have reappeared without any record of provenance. In the late 1980s a $5\frac{1}{2}$-inch, in its iron cell, was bought for £1 at a car-boot sale (the innocent vendor describing it as 'no good for shaving'). It is signed and dated 1863, and is therefore one of the earliest efforts of its maker – and it might well be the same $5\frac{1}{2}$-inch mirror that Webb tried before he purchased his 8-inch in 1864.

$9\frac{1}{3}$-inch With–Berthon reflector

In 1866 Webb's father (then aged 90) bought his son a new telescope, with a $9\frac{1}{3}$-inch silver-on-glass mirror by With and an 'equestrian' mount manufactured by E.L. Berthon.[38] Until that time, Webb had always observed in the open air, but with the acquisition of the $9\frac{1}{3}$-inch – the largest that he owned – he was prompted to build a small observatory. It was of the 'Romsey' type (also designed by E.L. Berthon, and named after his parish) – made of wood and canvas – and stood a few yards south-east of the front door of the vicarage. This new design[39] proved very popular with many amateurs, as it was easily adaptable and could be built at a cost of around £10.[40] According to Webb it was 'a very simple and cheap telescope-house, combining shelter with open-air freedom, to the great merit of which I can bear full testimony'.[41]

The $9\frac{1}{3}$-inch reflector was Webb's last telescope. The observatory proved of great benefit for many years, but it eventually began to deteriorate. On 2 February 1883 he told Ranyard:

> My telescope roof is all to pieces. I'll put on another roof with two pairs of opposite shutters, not only saving time in turning, but as the one side rises more steeply than the other, giving relief in position where now it cramps the head awkwardly. There was a bright thought.[42]

It may well be, however, that the repairs were never undertaken, as his last observation – on 19 March 1885 – was terminated when his candle was blown out by the wind.

Other instruments

Throughout his life Webb visited private observatories equipped with telescopes much larger than his own; but it was not in his nature to undertake the work to which these instruments were committed, although observations made during some of these visits are recorded in his journals.

In 1849 he travelled to London, and in late November and early December spent several evenings in the company of John Russell Hind, observing with the 7-inch Dollond refractor at George Bishop's observatory in the Inner Circle of Regent's Park (which at that time lay on the northern outskirt of the city). Bishop had established his observatory in the late 1830s, after amassing a considerable fortune in the wine trade; but he was not an observer, and he employed a succession of 'assistants' who carried out all of the observations. The first of these was W.R. Dawes, whose primary work was the micrometric measurement of double stars. Dawes was succeeded by Norman R. Pogson (who later became Director of Madras Observatory), W. Ellis, and Hind, whose reputation was formed chiefly by his discoveries of asteroids. Webb, however, observed stars and nebulae during his visit, and it may have been Hind's discovery of Nova Ophiuchi 1848 which attracted him.

Webb's visits to other observatories reveal that although he was content to observe with a 'common telescope', he was also keen to acquaint himself with the leading edge of research. On 7 December 1855 he observed Saturn with Warren De La Rue's 13-inch 10-foot focus reflector at Canonbury, London. De La Rue had made several 13-inch mirrors specifically worked for superior definition of planets, and in 1852 had begun experiments in astronomical photography with the newly invented wet-plate process. After his removal to Cranford, Middlesex, and his subsequent production of outstanding photographs of the Sun and Moon, his reputation was secure, and he was later commissioned to design the Kew photoheliograph for daily solar photography. Webb spent time with this pioneer, but he seems never to have been inclined to attach a camera to his telescope (although he and his wife possessed photographic equipment, and Mrs Webb was a keen photographer).

On 25 October 1872 Webb spent 'an evening never to be forgotten' at the observatory of William Huggins, at Tulse Hill, south London. Huggins had set up his observatory in 1856, and for two or three years had observed first with a 5-inch Dollond refractor, and afterwards with a Cooke refractor fitted with the 8-inch Clark object-glass which he had purchased from W.R. Dawes in 1858. He became dissatisfied with routine work, however, and in 1859 the news of Gustav Kirchhoff's revolutionary work in spectroscopy came to him 'like the coming upon a spring of water in a dry and thirsty land'. He soon afterwards began work on the analysis of laboratory spectra for comparison with stellar spectra, and in 1864 his studies of diffuse and planetary nebulae led to his discovery of gaseous spectra and the solution of 'the riddle of the nebulae' – that not all nebulae are resolveable into stars, and are intrinsically gaseous. Huggins' subsequent work secured his reputation as a pioneer in astrophysics – the new science – and in 1870 the Royal Society loaned him a 15-inch Grubb refractor and an 18-inch Cassegrain reflector (with speculum-metal mirrors). During his visit to Huggins' observatory Webb observed Neptune 'with the magnificent 15in O.G. and low power, the air not being good. The planet was a brilliant object ... of a bluish light, & showing a considerable disc.' He also observed several planetary nebulae, including the Saturn Nebula (NGC 7009, in Aquarius) with 'a very thin ray on each side', and the Ring Nebula (M57, in Lyra), of which he made small and delicate drawings, including a drawing of the spectrum of the Ring Nebula. His time with Huggins obviously inspired him, as six weeks later, on 5 December, he turned his new miniature direct-vision spectroscope, made by John Browning, onto the Great Nebula in Orion. This 'first trial', however, is the only record of this instrument. Spectroscopy, it seems, did not fit with Webb's contemplatory ethos.

Having observed with these large telescopes, with which he was obviously impressed, Webb must have had at least some desire to own such an instrument himself; and yet for the last two decades of his life he was perfectly content to use his 9⅓-inch With–Berthon reflector, which, although beyond his definition of a 'common telescope', always served his needs 'till the dappled dawn doth rise'.

Notes and references

1 According to Webb's friend and executor T.H.E.C. Espin, in his 'Reminiscence' in the later editions of *Celestial Objects for Common Telescopes*, Webb's earliest recorded manuscript observation (of Jupiter) was made with a 3.7-inch Tulley refractor on 3 July 1834. This, however, was the first observation recorded in his observing journals, whereas the earlier observations were recorded in his diary. The 'Reminiscence' also includes several other errors.

2 The results were detailed in a paper read before the Royal Society of Edinburgh, 3 and 4 April 1791.

3 *Philosophical Transactions of the Royal Society*, **118** (1828), 105; **119** (1829), 33–46; **121** (1831), 9–15; **123** (1833), 1–13.

4 Robert Brettell Bate (1782–1847) was mathematical instrument-maker to

HM Excise, Optician in Ordinary to George IV, William IV and Queen Victoria, and Master of the Spectaclemakers' Company. His output included mathematical, scientific and optical instruments, hydrometers, standard weights and measures, bullion balances, and books on navigation, but the telescopes he produced are now seldom encountered.

5 T.W. Webb, 'The Achromatic Telescope, Dialytes, and Fluid Lenses', *Intellectual Observer*, **7** (1865), 179–90.

6 A. Rogers, 'On the Construction of Large Achromatic Telescopes', *Monthly Notices of the Royal Astronomical Society*, **1** (1827), 71.

7 In chemistry, dialysis is the process of separating the soluble crystalloid substances in a mixture from the colloid by means of a dialyser – a vessel formed of parchment or animal membrane floated on water, through which the crystalloids pass, leaving the colloids behind.

8 Letter to John Webb, April 1832, in private collection.

9 Webb (n. 5).

10 J. Wall, 'Building a 30-inch refractor', *Journal of the British Astronomical Association*, **112** (2002), 260. The plate glass for the 30-inch lens cost £100. This instrument is now at the Hanwell Observatory, near Banbury.

11 W. Kitchiner, *The Economy of the Eyes*, part 2 (1818), p. 28.

12 W.H. Wollaston (1766–1828) was a Fellow of the Royal Society, served as Secretary and President, and was recipient of the Copley Medal and the Royal Medal. In 1789 he settled as a physician in Bury St Edmunds. His contributions to the *Philosophical Transactions of the Royal Society* included papers on astronomy, optics, chemistry, mechanics, acoustics, mineralogy, crystallography, physiology, pathology and botany. In 1828 he self-diagnosed a brain tumour and, certain of his imminent demise, presented his Dollond telescope to the Astronomical Society of London, and gave £2,000 Consols to the Royal Society.

13 Letter from W.H. Wollaston to J.F.W. Herschel, *Monthly Notices of the Royal Astronomical Society*, **1** (1829), 103.

14 RAS Letters, Webb, 27 April 1852.

15 RAS Letters, Webb, 9 June 1857.

16 For a detailed description of this instrument and its optical system, see H.E. Dall, E.J. Hysom and C.A. Ronan, 'A Dollond–Wollaston Telescope', *Journal of the British Astronomical Association*, **90** (1980), 422–8.

17 This was a very large sum of money. As a comparison: when seeking a new property a few years earlier Dawes had told John Herschel that he wanted a large house with 2 or 3 acres of grounds, for which he was prepared to pay £40 a year. Shortly after acquiring the $8\frac{1}{4}$-inch object-glass, Dawes commissioned Clark to make a tube and mount for it, and the instrument thereby became the first complete Clark refractor outside America. In 1870 it passed to Rugby School, where it is still in use. For a detailed account, see R.A. Marriott, 'The $8\frac{1}{4}$-inch Clark refractor of the Temple Observatory, Rugby', *Journal of the British Astronomical Association*, **101** (1991), 343–50.

18 RAS MSS Webb 5.

19 *Memoirs of the Royal Astronomical Society*, **30** (1862), 65–8.

20 Webb to Ranyard, 12 August 1859.
21 Sir David Gill (1843–1914) was HM Astronomer at the Cape of Good Hope, 1879–1907.
22 The Revd Edward Lyon Berthon (1813–1899) was vicar of Holy Trinity, Fareham, 1846–59, and of Romsey, 1859–92. Besides theology, he excelled in medicine, architecture, philosophy, mechanics, sculpture and art. He invented and designed several nautical devices (including the collapsible boat), and for astronomy he produced the 'Romsey' observatory, several types of telescope mount, and 'Berthon's dynamometer' – a device for measuring the exit pupil of an eyepiece.
23 H. Cooper Key, 'On a mode of Figuring Glass Specula for the Newtonian Telescope', *Monthly Notices of the Royal Astronomical Society*, **23** (1863), 199–202.
24 R.A. Marriott, 'The life and legacy of G. H. With, 1827–1904', *Journal of the British Astronomical Association*, **106** (1996), 257–64.
25 Cooper Key (n. 23).
26 RAS Letters, Webb, August 1863.
27 See G. Parry Jenkins, 'A plea for the reflecting telescope', *Journal of the Royal Astronomical Society of Canada* (1911), 59. Parry Jenkins was a friend of With, and had lived in Hereford for ten years before emigrating to Canada.
28 Webb to Ranyard, 25 November 1863.
29 Cooper Key (n. 23).
30 Webb to Ranyard, 25 November 1863.
31 Cooper Key (n. 23).
32 T.W. Webb, *Celestial Objects for Common Telescopes* (later editions).
33 Webb to Ranyard, 8 April 1875.
34 Webb to Ranyard, 16 March 1880.
35 R.J. McKim, 'Nathaniel Everett Green: artist and astronomer', *Journal of the British Astronomical Association*, **114** (2004), 13.
36 H.E. Dall, 'George Calver: East Anglian telescope maker', *Journal of the British Astronomical Association*, **86** (1975), 49. By about 1870 Calver had moved to Widford, near Chelmsford, Essex, and in 1904 he returned to Walpole (his birth-place). Dall corresponded with him in 1923.
37 Espin was vicar of Tow Law, County Durham, from 1888 to 1934. The author's paternal grandfather was born in Tow Law around 1890, and lived there until active service during the First World War; but whether he was baptised by Espin has yet to be ascertained.
38 In 1943, C. Waller presented the British Astronomical Association with a With–Berthon reflector (see illustration, p. 128). Over the years it came to be believed that it had been Webb's telescope, and during the late 1970s D.G. Buczynski carried out a large amount of research to determine whether this assertion was true, as many features of the instrument seemed to correspond. The mirror is inscribed 'Withus Herefordensis me ad astra investiganda fecit – Laus Deo'; but unfortunately, and uncharacteristically, With did not date it. It seemed that the question would remain unanswered. However, the author of this chapter has more recently read

through Webb's journals (which were not readily available to Buczynski) and made precise measurements of the mirror. It is exactly $9\frac{1}{4}$ inches in diameter, whereas Webb initially refers to his telescope as a 9-inch, and later a $9\frac{1}{3}$-inch, a 9.38-inch, and even a 9.368-inch. It can only be concluded, therefore, that it is not the same mirror, and that Webb's telescope passed into obscurity and may well no longer exist.

39 Described in *English Mechanic* (1871).

40 A description of the building of a Romsey observatory, together with folding plans for two different sizes, was published by G.F. Chambers in the second volume of *A Handbook of Descriptive and Practical Astronomy* (Clarendon Press, Oxford, 1890).

41 Webb (n. 32).

42 Webb to Ranyard, 2 February 1883.

Chapter 9

Webb's observing notebooks

Peter D. Hingley

Do not avoid the trouble of recording regularly all that you see, under the impression that it is of no use. If it has no other good effect, it tends to form a valuable habit of accuracy; and you might find it of unexpected importance.

The Webb manuscripts were donated to the Royal Astronomical Society in 1978, through the good offices of Dr D.W. Dewhirst – of the Institute of Astronomy, Cambridge, and for many years the Chairman of the RAS Library Committee – who had been approached by Mr J. Sutherland of Newcastle, on behalf of a Mrs Milburn. Dr Dewhirst encouraged Mrs Milburn to deposit the manuscripts with the Society, and they have been in the archives since then. The lady concerned was the widow of William Milburn, assistant to the Revd Thomas Espin at Tow Law, County Durham, from 1912 until Espin's death in 1934. Without Milburn's recognition of the importance of the notebooks they might well have perished, for when Espin died the bonfire of papers taken from his vicarage burned continuously for three days.[1]

The following concentrates on the physical description of the books. One volume has partly original pagination by Webb himself, and the rest of this volume and the other volumes were paginated by the present writer in 1990. It is immediately obvious that these volumes must be 'fair copies' for a permanent record, as surely even the meticulous T.W. Webb could not have produced such neat and even script, in his usual well-rounded hand, without the crossings-out endemic in books used for practical observation! Indeed, Espin states that 'the rough notes were made at the telescope, prefaced by the state of the

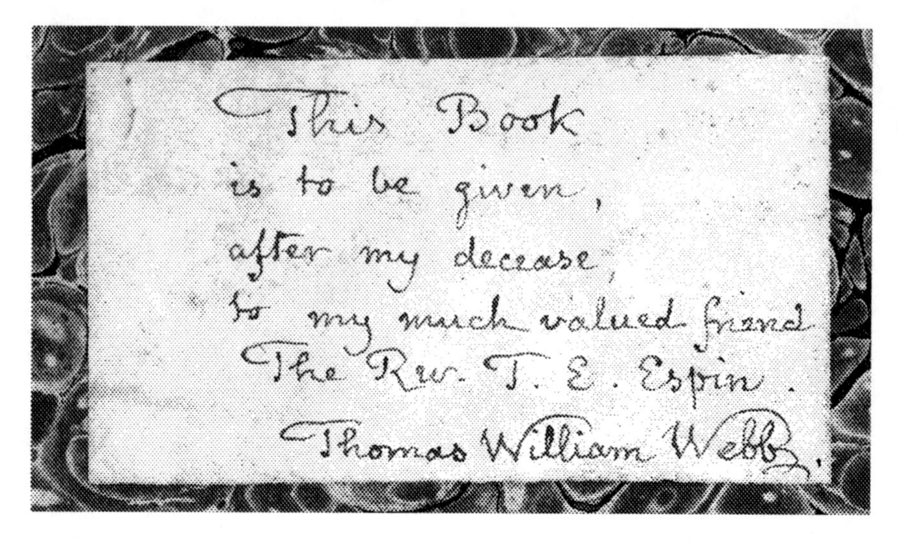
This Book
is to be given,
after my decease,
to my much valued friend
The Rev. T. E. Espin.
Thomas William Webb.

The inscription inside the front cover of Webb's notebook, Volume 5.

air, the powers used etc. and written out in full the next day'. Webb was evidently concerned that his records should be preserved – as can be seen from the note inside the front cover of Volume 5 (above).

RAS MS Webb 1

'Abstract of the Bedford Catalogue, forming Vol. II of "A Cycle of Celestial Objects" by Captain William Henry Smyth, R.N., with corresponding observations by Thomas William Webb' ΔΟΞΑ ΤΩι ΘΕΩι.

8 inches high x 10 inches wide; 150 pages, written throughout on feint lined paper; cover worn, purple-brown leather with, in the centre, an embossed block with the initials 'T.W.W.'

Inside front cover inscription, in Webb's hand: 'Thomas William Webb, the Gift of his Father, December 14, 1847.' On the leaf facing the 'title page' Webb carefully notes abbreviations for the sources of observations and the use of Piazzi's magnitudes. He notes a few inaccuracies in Smyth's positions (after Smyth's death there was a considerable controversy about his measurements), and also that Smyth's colour estimates – some of which, to modern eyes, seem rather exaggerated – were in some cases confirmed by his great friend Dr John Lee, of Hartwell, Buckinghamshire (where Smyth latterly observed), while in others 'the eyes of ladies were consulted'. A pencil note states that Webb used his 3.7-inch Tulley refractor for these observations. Stuck to the rear endpaper of this book are sketches of stars in the area of Webb's new variable star S Orionis, dated 1871, by his fellow observers J. Baxendell and G. Knott.

It is interesting to note that Webb gives Smyth's rank as Captain. This famous amateur astronomer reached this rank in active service, but on his retirement in 1829 he was promoted to Admiral, and is usually known as Admiral Smyth (1788–1865). His *Cycle of Celestial Objects*, published in 1844, gives his rank as Captain. His friends called him 'The Admiral'.

RAS MS Webb 2

'Moon'

Worn soft-covered notebook with purple–brown cover, and title written on small paper label on front cover; 8¾ inches high x 7 inches wide; 53 pages used, remainder blank.

Dated observations of the Moon are recorded between 15 June 1864 and 2 February 1884. Many small and detailed drawings of the Moon – made with a fine pen in Webb's characteristic 'miniature' style – are in a few cases on tracing paper stuck in, and were presumably for the transfer of drawings from a rough observing book used at the telescope.

RAS MS Webb 3

'Observations on Double Stars and Nebulae by T W Webb'

Reasonably sound exercise book, lined paper, with black cover; 9 inches high x 7½ inches wide; 101 pages used, remainder blank.

Inscription inside front cover, in an italic hand, very unlike TWW's own, so presumably that of Henrietta: 'Thomas William Webb, from his affectionate wife, St Matthews 1871, Hardwick.' A further tracing of stars around S Orionis is stuck to the flyleaf, by H. Sadler, 1876. (Sadler was the young man who, a couple of years later, intemperately and controversially attacked the accuracy of many of Admiral Smyth's measurements.[2])

RAS MS Webb 4

'Miscellaneous Observations by T.W. Webb 1874'

Small and rather fragile sewn notebook, card covers; 7 inches high x 4¾ inches wide; 20 pages used, remainder blank.

After the black ink of the title Webb has added in red: 'N.B. There are more note worthy [*sic*] observations in this little book than might be inferred from its appearance.' There are indeed some noteworthy observations of happenings in the Border country. The majority of these are tremors felt by Webb and others, and these have been excellently interpreted by Dr Musson (in Chapter 6). Webb was eager to include tremors noted by other people. For example: on 19 November 1879, Thomas Espin was visiting the Webbs and felt a vibration which was duly noted; on 27 January 1883 Mr De La Touche, the vicar of Stokesay, felt a vibration which was subsequently reported in the *Guardian*; and so on. Reports of other phenomena include notes of spectacular aurorae, zodiacal light, and in large letters, BLACK RAIN, noted in May 1880 and May 1882.[3]

RAS MS Webb 5

'Astronomical Observations made by Thomas William Webb'
ΔΟΞΑ ΘΕΩ.

Worn exercise book bound in what was originally green half leather; 8 inches high x 7 inches wide; paginated in red ink, by Webb, from the beginning up to p. 119, with the modern pagination continuing to p. 297. Webb has used the end of the book, upside down, paginated 1–56 by himself.

This is by far the most evocative of Webb's observing books, as it contains sections with observations and drawings of various types of celestial object in several sections. In some cases, letters and newspaper cuttings relevant to the objects described are stuck in. Some of the drawings are in plain pencil or ink, and others are delicately heightened with coloured crayons or watercolour. In most cases the drawings are on Webb's characteristically small scale, with planetary discs rendered ½-inch diameter or less, although the giant Jupiter sometimes reaches 1¼ inches! It seems that Webb divided up this book in advance to take his fair copy observations, but sometimes ran out of pages.

Stuck in the front is the portrait of Prebendary and Mrs Webb (see frontispiece).

Pages 1–47 contain 'Observations on Jupiter' dated between 2 December 1834 and 30 December 1870. Pages 48–64 have observations of Mars between 8 March 1839 and 11 January 1865, and at the end of this section Webb notes that observations of the planet were continued in 1871, in a separate book. Pages 66–72 are occupied by Venus, with some delicate drawings of the phases, but without too much of the very indistinct surface detail. On pp. 73–7 Jupiter returns, but the observations, covering 29 August 1879 to 20 February 1885, are out of chronological sequence. Pages 81–8 cover 30 December 1870 to 6 November 1872. What was often called 'our great central Luminary', the Sun, then appears, with pp. 90–6 bearing observations from 29 June 1847 to 22 August 1862. There is then a record of the transits of Mercury on 9 November 1848 and 12 November 1861. Webb was disappointed by the annular eclipse of 15 March 1858, but saw the total eclipse of 22 December 1870 reasonably well, despite intermittent cloud. The magnitude at Hardwicke (where it was partial) he records as 0.8. (To observe this eclipse, various official and private British and American observing parties travelled to Cadiz, Spain, and Oran, Algeria.) Pages 188–214 contain one of the most evocative sections: comets – the observations of which, covering 11 June 1845 to August 1862, are also out of sequence. Pages 96–104, covering 18 August 1862 to 14 November 1882, are graced with some of Webb's most beautiful and evocative drawings (see pp. 186, 189 and 191). There are two images which very strikingly epitomise the frustrations of the practical observer and the eternal battle between author and publisher. On p. 103 is a small wash drawing of the tail of the Great Comet of 1882, the head of which is not included, because on 3 October it had set behind the trees at Hardwicke before it could be observed – probably due to bad weather, although Webb adds that he saw it again from Cheltenham on 25 October. On p. 88 is a series of beautiful and delicate drawings of Donati's comet, with what must have been the largest cut-out, leaving a large hole in the page, patched with a piece of paper on which Webb has written:

> NB this space contained my sketches of Donati's Comet, which were cut out and sent to the engraver in London, for the publication of my Observations in the Memoirs of the Royal Astronomical Society. The Engraver was requested to return them; but it was never done. The engravings represent them very fairly.

Anyone who has ever lost anything at the hands of a printer will feel for Webb at this point!

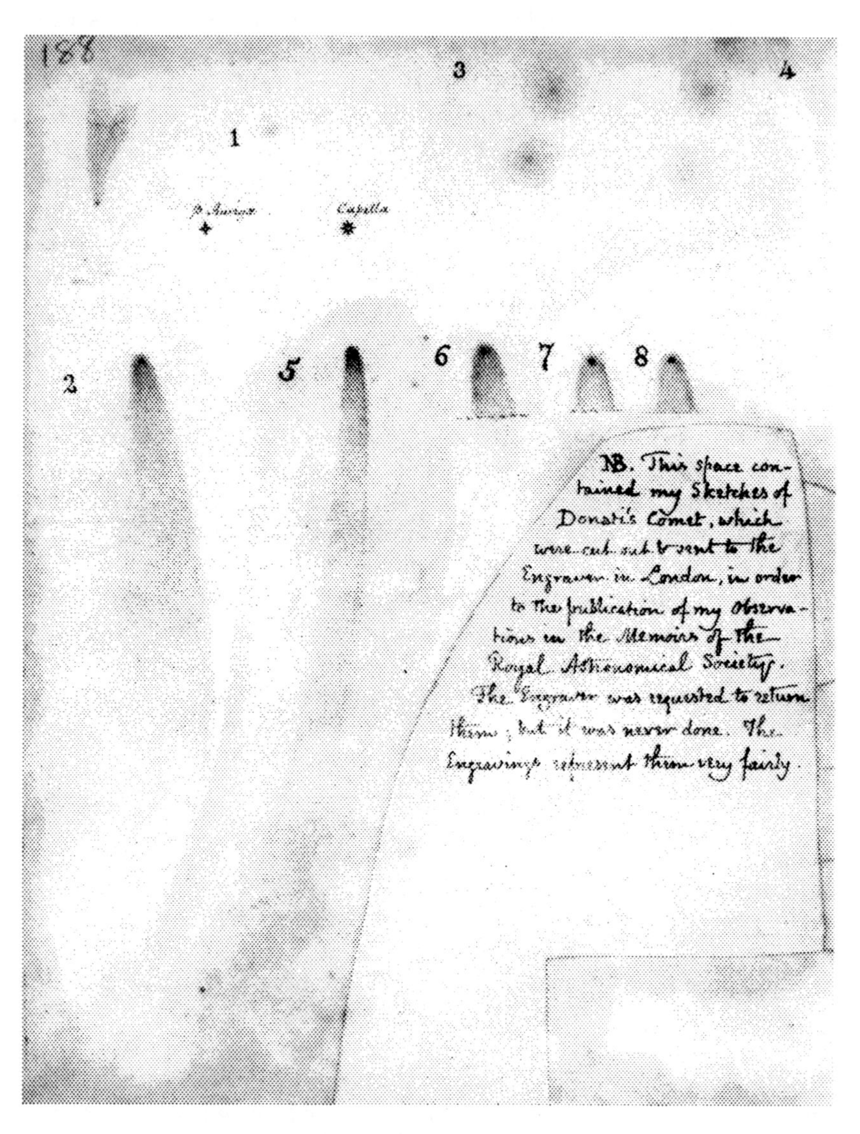

Donati's comet, 1858, in Webb's notebook, Volume 5.

The sidereal heavens make cameo appearances in this book, although Webb's main records of stellar observations are elsewhere: Sirius on p. 109, Alpha Orionis on p. 110, Stars in Cassiopea on p. 112, New Star in Ophiuchus on p. 114, and Corvus and Crater on p. 116. Minor planets – still then a field of exciting discovery and not yet 'the vermin of the Heavens' – with Melpomene, Vesta, Ceres and Pallas, are on pp. 118–20. Jupiter returns on pp. 110–11, 113 and 115, and there are a few observations of Uranus on pp. 117 and 121 and of remote Neptune on p. 122. Ever magnificent Saturn appears on pp. 124–34 with a detailed drawing of the 'Ring'; and then on to one of Webb's favourite objects, the Moon. These observations are 'Continued from an old book in a marble cover', and include a number of small drawings of surface details, with many careful notes. Stars and nebulae appear again on p. 215 to the end of the front section on p. 296. The 'reversed' section at the end of the book mainly consists of transcriptions of historical observations, particularly by Herschel, classical quotations, and so on.

The notebooks can be summarised by quoting T.H.E.C. Espin in his 'Reminiscence' of Webb in the fifth edition of *Celestial Objects*:

> The observations are interspersed with numerous exquisite drawings. The manuscript observations are a model of neatness, patience and care.

Notes and references

1 Told to the Robinsons by the vicar of Tow Law in 1993.
2 J.L.E. Dreyer and H.H. Turner (eds.), *History of the Royal Astronomical Society, 1820–1920* (RAS, 1923).
3 See Chapter 6, n. 2.

Chapter 10

Webb and the Moon

William Sheehan

We shall find the Moon a wonderful object of study. It presents to us a surface convulsed, upturned, and desolated by forces of the highest activity, the results of whose earliest outbreaks remain, not like those of the Earth ... comparatively undegraded from their primitive sharpness even to the present hour.

Thomas William Webb was many things: priest, humanitarian, astronomer. Above all, he was a man whose highest court of appeals was the eye – possessed with his insatiable desire 'to see'. Like his younger contemporary John Ruskin he was dominated by the sense of vision, occupied with being witness of everything that passed overhead in the grand pageant of clouds and sky.

That the heavens were a branch of natural history – thus capable of close observation and useful classification – was a notion pioneered by William Herschel in his great studies of stars and nebulae. Webb's interest in all aspects of the natural world was for him of one piece; they reflected his sense of the Creation as divine. One of his favourite expressions, often repeated, was that of the German astronomer Johann Hieronymus Schröter (1745–1816), who reflected on the unity of the diversity in the natural world – the common purpose and harmony of pattern underlying the many individual types. The design was simple and unitary; the details multitudinous and complex. It was this reinforcing pattern that was the strongest testimony of its divine construction.

Webb felt a sense of kinship with Schröter. Both were involved in careers of service: Webb as a curate, who chose the wider margins of rural life and indulgence of a vocational interest over the allures of the lively but less contemplative Metropolis; Schröter as a lawyer who might have advanced at the court of Hanover but instead retreated to the slower rounds of the 'Vale of the Lilies', where he could pursue the subject that really interested him. Although Webb spent much time wandering among the starry paths, his real interest was, like Schröter's, in the Solar System. He was not interested in the problems of celestial mechanics – the professional preoccupation of the eighteenth and early nineteenth centuries – but was instead a pioneer of selenology and planetology, the understanding of other worlds, their make-up as canopies of cloud or surfaces of rock.

Although Webb's observations of the planets may have been, statistically, more numerous than any other class of data entered into his observing diaries, his most detailed studies were of the lunar surface. He was, like Schröter, and many a keen amateur since, first and foremost a 'Moon man'. In *Celestial Objects for Common Telescopes* – his masterpiece, the distillation of a lifetime of nightly vigils – the chapter on which the most care is lavished and which shows the widest reading and the profoundest reflection and meditation is the chapter on the Moon. Here Webb shows a keen familiarity with the quintessential science of the first half of the nineteenth century: geology. He knew that the Moon was, based on its proximity and susceptibility to detailed study and the analogy of its features, capable of being investigated in geological terms. He was a backyard prospector – in effect, a telescopic field geologist.

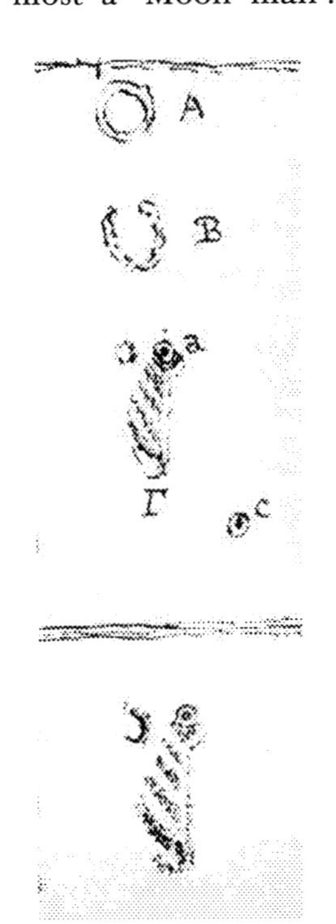

The crater Schröter, in Webb's notebook, Volume 2.

He took inspiration from Schröter, whose quest to establish the reality of lunar change motivated him; also – though more by reaction – from Beer and Mädler's thesis, published in 1837, that the Moon was 'no copy of the Earth', but a static and immobile plaster-of-Paris mask, a mask of death.

Among contemporary lunarians, Webb became a follower of the visionary British geologist John Phillips. Phillips proposed the systematic study of selected regions of the Moon – a study that would be analogous to the geological fieldwork long pursued by amateurs of science, whose familiarity with local terrain gave them

> ... a positive advantage over the metropolitan or university geologists who might occasionally visit their home areas. Within those areas, amateur geologists had the time that was needed to find fossils by patient hammering in local quarries, cliffs or cuttings, for fossils were rarely to be found without such time-consuming work.[1]

For that matter, geology's popularity among amateur scientists of the nineteenth century has been explained, in part, by the sheer accessibility of the subject, since little more was required than a hammer, a magnetic compass, a clinometer for measuring the tilt or dip of strata, and perhaps a hand lens. Lunar observation was similarly accessible. One need not employ mammoth instruments like the reflectors of the Herschels or Lord Rosse. Apart from a modest telescope, one needed only diligence, a pencil and a sketchpad.

There were romantic aspects to both geological fieldwork and astronomical observation. In both disciplines, as Martin J.S. Rudwick has said of geology, 'the primacy of fieldwork reflected a recognition that the empirical locus of a science like geology was bound to lie "in the field"'.[2] Amateurs could therefore contribute more substantially to geology and astronomy than to most other sciences. There would long remain a distinctly Baconian flavour to these investigations, by which the collection of facts was generally preferred to far-flung theoretical speculations.

Webb's own descriptions of the Moon's individual features, recorded in the notebooks preserved in the Royal Astronomical Society's archives, are careful, even loving. For example:

> Oct. 19, 1855. Clavius a magnificent object, with its included craters and monstrous terraced ring, casting a vast shadow into the depth below ... Oct. 23, 1855. Dörfel or some other mountains in grand and lofty projection, pointed at from Tycho through W. Wall of Clavius, and consequently near S. Pole ... Wargentin a curious object – somewhat like a thin dark cheese ... Nov. 1, 1855. Maginus with very remarkable serpentine terraces in the inside of its W. wall, and a number of craters which have reduced the E. wall to the condition of a gigantic ruin.

It was, of course, in the collection of such facts and descriptions that amateurs excelled. They were the amateur naturalists, the beetle-collectors and butterfly-chasers of science – precisely the kind of men and women who could, with the proper direction, undertake the important role of carrying out the arduous and painstaking fieldwork at local quarries and cliffs or the comparable surveillance of the minor surface features of the Moon.

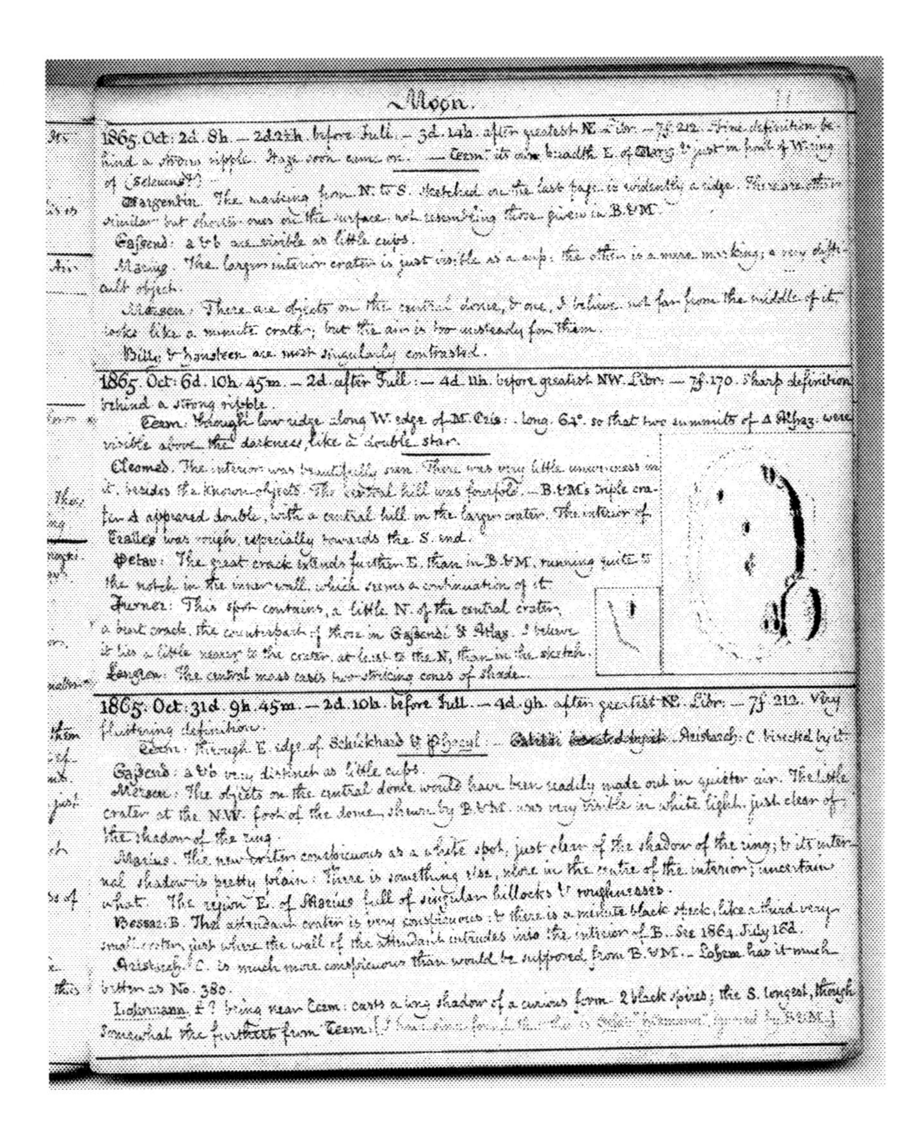

Moon.

1865. Oct: 2d. 8h. — 2d.22h. before Full. — 3d. 14h. after greatest N. [illegible] — 7f. 212. Fine definition behind a [illegible] ripple. Haze soon came on. — Term: its [illegible] breadth E. of [illegible] & just in front of W. [illegible] of (Seleucus?) —

Margentin. The markings from N. to S. sketched on the last page is evidently a ridge. There are other similar but shorter ones on the surface not resembling those given in B.&M.

Gassend: a & b are visible as little cups.

Marius. The larger interior crater is just visible as a cup: the other is a mere marking; a very difficult object.

Mersen. There are objects on the central dome, & one, I believe, not far from the middle of it, looks like a minute crater; but the air is too unsteady for them.

Billy & Hansteen are most singularly contrasted.

1865. Oct: 6d. 10h. 45m. — 2d. after Full: — 4d. 11h. before greatest NW. Libr. — 7f. 170. Sharp definition behind a strong ripple.

Term. through low ridge along W. edge of M. Cris: long. 64°. so that two summits of A Alhaz. were visible above the darkness, like a double star.

Cleomed. The interior was beautifully seen. There was very little unevenness in it, besides the known objects. The central hill was fourfold. — B.&M's triple crater A appeared double, with a central hill in the larger crater. The interior of Tralles was rough, especially towards the S. end.

Petav: The great crack extends further E. than in B.&M. running quite to the notch in the inner wall, which seems a continuation of it.

Furner: This spot contains, a little N. of the central crater, a bent crack, the counterpart of those in Gassendi & Atlas. I believe it lies a little nearer to the crater, at least to the N, than in the sketch.

Langren. The central mass casts two striking cones of shade.

1865. Oct: 31d. 9h. 45m. — 2d. 10h. before Full. — 4d. 9h. after greatest NE. Libr. — 7f. 212. Very fluttering definition.

Term. through E. edge of Schickhard & Phocyl. — [illegible] Aristarch: C bisected by it.

Gassend: a & b very distinct as little cups.

Mersen: The objects on the central dome would have been readily made out in quieter air. The little crater at the N.W. foot of the dome, shown by B.&M. was very visible in white light, just clear of the shadow of the ring.

Marius. The new crater conspicuous as a white spot, just clear of the shadow of the ring; & its internal shadow is pretty plain. There is something else, more in the centre of the interior; uncertain what. The region E. of Marius full of singular hillocks & roughnesses.

Bessar: B. The attendant crater is very conspicuous: & there is a minute black speck, like a third very small crater, just where the wall of the attendant intrudes into the interior of B. See 1864. July 16d.

Aristarch: C. is much more conspicuous than would be supposed from B.&M. — Lohrm has it much better as No. 380.

Lohrmann. 1? being near Term: casts a long shadow of a curious form - 2 black spires; the S. longest, though somewhat the furthest from Term. [illegible]

Webb's notebook, Volume 2, p. 11.

Inevitably, the Baconian approach was problematical, for the observations themselves were often theory-laden. The situation was no different from that which Charles Darwin noted in his ramblings with Cambridge geologist Adam Sedgwick among the Cambrian rocks, during which he proved oblivious to the abundant evidence of past glacial action all around him. One saw as much – and only as much – as one was prepared to see. As Darwin later wrote:

> On this tour I had a striking instance how easy it is to overlook phenomena, however conspicuous, before they have been observed by anyone. We spent many hours at Cwm Idwal, examining all the rocks with extreme care as Sedgwick was anxious to find fossils in them; but neither of us saw a trace of the wonderful glacial phenomena all around us; we did not notice the plainly scored rocks, the perched boulders, the lateral and terminal moraines. Yet these phenomena are so distinct that ... a house burned down by fire did not tell its story more plainly than did this valley.[3]

In this respect it is important to note the theory, or at least the assumption, that lay in the background of many of the leading amateur observers of the Moon: that of ongoing change – the Moon was not dead; it was still the scene of atmospheric obscurations and perhaps even of occasional volcanic eruptions.

The romantic appeal of exploring the remote quarries and cliffs of the Moon – of wandering, visually, over fields that presented objects as impressive as, and possibly analogous to, the extinct volcano fields of the Auvergne – was not lost among observers of that distant orb. Its features were charged with interest. As Webb pointed out in *Celestial Objects*, the Moon presented

> ... a surface convulsed, upturned, and desolated by forces of the highest activity, the results of whose earliest outbreaks remain, not like those of the Earth, levelled by the fury of tempests, and smoothed by the flow of waters, but comparatively undegraded from their primitive sharpness even to the present hour. The ruggedness of the details, as old Hevelius had anticipated, becomes more evident with each increase of optical power, and we cannot doubt that we look upon the unchanged results of those gigantic operations which have stereotyped their record on nearly every region of the lunar globe.[4]

Proficient in German, Webb read Beer and Mädler in the original language, and became the most reliable source of information about German selenographic studies, then the standard, in the English-speaking world. At Gloucester he began routine observations of the

Sun, Moon and planets with his 3.7-inch Tulley refractor, and immediately realised that there was still much useful work to be done, especially on the Moon:

> Lohrmann, Beer and Mädler have done admirably well in their delineations. Yet a little experience will show that they have not represented all that may sometimes be seen with a good common telescope. My own opportunities, even when limited to a $3\frac{7}{10}$-in. aperture, satisfied me, not only how much remains to be done, but how much a little willing perseverance might do.[5]

The lengthy chapter on the Moon presented a complete – if, for the time, conventional – view of the lunar surface. The 'crater mountains' were 'the grand peculiarities of the Moon: commonly, and probably with correctness, ascribed to volcanic agency'.[6] Webb mentioned their resemblance to rings left by gaseous bubbles, first ingeniously invoked by Hooke, only to dismiss the analogy on the grounds that no materials could have sufficient cohesiveness to support structures on such an enormous scale. Instead he preferred a modified bubble theory – the idea that the lunar craters were the remains of molten lakes on a once-fiery globe.

What fascinated Webb above all else was the exciting possibility that the surface of the Moon was still in active condition. His first essay, 'On the Lunar Volcanoes', was presented to the British Association for the Advancement of Science as far back as 1838, but not until 1859 did he follow it up with 'Notice of Traces of Eruptive Action in the Moon'. Here he pointed out that while astronomers were generally agreed as to the cessation of such action on a large scale, 'this would not necessarily imply the impossibility, or even improbability, of minor eruptions, which might still continue to result from a diminished but not wholly extinguished force'.[7] He thus reintroduced the Schröterian project of documenting lunar changes, which had been all but banished ever since Beer and Mädler published their classic negative verdict in 1837.

Webb recommended two specific regions for further detailed study. The first was the crater Cichus – a rather obscure feature in southern Mare Nubium. It seemed to have grown larger compared to the earlier representations by Schröter and Mädler. This was an admittedly weak case; but the other site was more convincing, and indeed would become one of the most scrutinised places on the Moon by students pursuing evidence of change.

In Mare Foecunditatis, the name Messier had been given to one of two small craters lying side by side at the end of parallel white streaks 'resembling the divided tail of a comet'. Schröter had drawn the region; and so had Beer and Mädler, who claimed to have inspected the formation no fewer than three hundred times between 1829 and 1837. They always found the two craters 'as like as two peas in a pod – perfectly and singularly alike in size, shape, height of ring, depth of cavity, and even in the position of some peaks upon the ring'. Nothing, apparently, could be more definite; but Webb, first on 14 November 1855, and on many occasions afterwards, found that whatever had been their previous resemblance, the two craters were no longer 'peas in a pod'. The difference between them was in fact so striking 'as to indicate a permanent alteration of the surface during the space of twenty years'.[8]

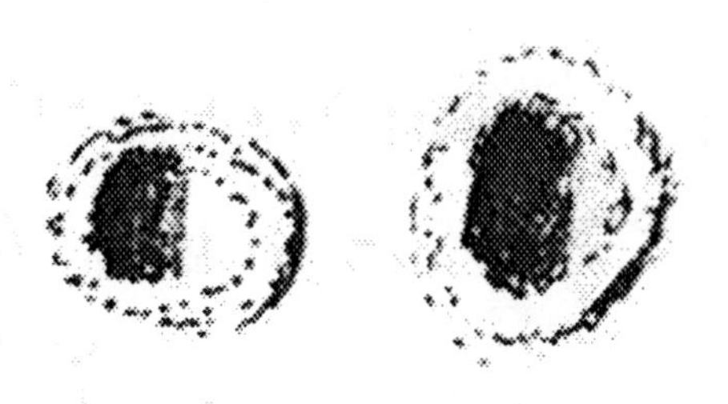

The twin craters Messier, in Webb's notebook, Volume 2.

Such comments show that seven years before the Linné affair, the notion of lunar change was once more in the air. Indeed, it is hardly surprising that astronomers had rebelled against – or at least found themselves unwilling to embrace – the static desert world of Beer and Mädler. As Richard Anthony Proctor would sum up a few eventful years later:

> The examination of mere peculiarities of physical condition is, after all, but barren labour, if it leads to no discovery of physical condition. The principal charm of astronomy, as indeed of all observational science, lies in the study of change – of progress, development, and decay, and specially of systematic variations taking place in regularly-recurring cycles ... In this relation the Moon has been a most disappointing object of astronomical observation. For two centuries and a half, her face has been scanned with the closest possible scrutiny; her features have been portrayed in elaborate maps; many an astronomer has given a large portion of his life to the work of examining craters, plains, mountains and valleys for the signs of change; but hitherto no certain evidence ... has been afforded that the Moon is other than 'a dead and useless waste of volcanoes'.[9]

The charm of geological ruins on the Moon – level wastelands, craggy wildernesses – was never so great as the earlier vision of a still-burning and active planet, with volcanoes, exhalations, erosive forces; perhaps even life of its own. So the romantic vision of the Moon died hard. It is this psychological fact which probably explains the mid-nineteenth century's growing preoccupation with apparent minor eruptions as evidence of the Moon's ongoing geological activity, along with the resurgence of the possibility that the Moon was a living world – 'habitable', as Webb expressed the hope, 'in some way of its own'.[10]

In March 1862 Phillips presented to the British Association, 'Suggestions for the Attainment of a Systematic Representation of the Physical Aspect of the Moon'. He began by pointing out that even the map of Beer and Mädler met the current needs of selenography only about as well as maps of England of the last century satisfied the requirements of physical geography. He called for renewed efforts at mapping the Moon's surface. 'In the same proportion as the great one-inch Ordnance Map of 1862 is superior to the old Chart of 1800,' he urged, 'so should be the new drawings of the features of the Moon to the older delineations.'[11]

Phillips had argued the need for such work to be accomplished with one instrument, which might be used in round robin fashion by several observers seriously interested in contributing to selenography. He offered his own 6-inch Cooke refractor to the cause in the hope that it might 'become the property of some scientific body constituted for long endurance'. In addition to his activity within the British Assocation, Phillips was also soliciting the support of the Royal Society. He submitted a request for a grant for the telescope to General Edward Sabine (1788–1883), Secretary of the Royal Society and Chairman of the Government Grant Committee. Sabine acknowledged 'the importance of giving a new and well considered impulse to Selenography', and recognised 'the value of your opinions on ... points of detail on which you have obviously thought so much'.[12] Nevertheless, the Committee declined Phillips' first request. Later it reversed itself and granted him the sum of £100 to help defray the costs of setting up his telescope in an observatory next to his new house in Oxford – a task completed in July 1862.

Phillips' initiative was taken up by William Ratcliff Birt (1804–1881), who emerged as the outstanding figure in British selenography of his era. From 1839 to 1843 Birt had served as John Herschel's assistant in reducing the latter's barometric observations made in various parts of the world. After 1843 Herschel withdrew, and Birt alone continued the investigation, publishing five reports on atmospheric pressure waves in the volumes of the British Association for 1844–48. In 1853 he also published a *Handbook on the Law of Storms*. However, from 1859 onward the pioneer meteorologist's interest lay almost entirely in the study of the surface of the Moon.

Above all, Birt shared Webb's burning interest in the question: Is the Moon's surface presently in an active or inactive condition? He admitted the possibility that all volcanic activity had long ago ceased – that the largest lunar forms had been the result of the most violent outbreaks, while the smaller mountains, especially in the larger craters, indicated 'the last expiring efforts of this action'. However, he added, the matter was far from settled, and 'astronomers were not wanting who leaned to the hypothesis that eruptive action still exists, although in a subdued form'.[13]

Birt himself was among the most hopeful of the hopeful astronomers who leaned strongly toward the hypothesis that eruptions still took place from time to time. He attempted to draw up a careful list of craters mapped by Schröter and others which had been missed by Beer and Mädler, 'the acknowledged authority in lunar matters', finding no fewer than 368 objects. Many were very small, and he readily admitted that while they raised the suspicion of change, 'our existing records were inadequate to determine the question'.[14]

Webb, too, compiled a list of features,[15] among which was a small pit, near the ring of Le Verrier (Beer and Mädler's Helicon A), which he described as 'extremely minute, about the minimum visible' in his notebook on 26 May 1836, but recorded as 'very conspicuous' on 20 April 1861. (By way of alternative explanation to an actual change, he conceded that he had by then graduated to a larger telescope.) Another of his suspect objects was 'a minute, but unquestionable, crater on the summit of the [western] wall of Helicon'; and yet another was a small crater in Mare Imbrium, 'more conspicuous than several small craters in the neighbourhood figured by Beer and Mädler'.

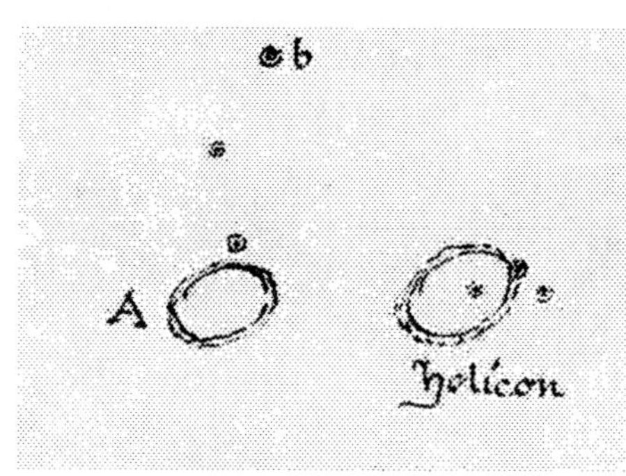

Helicon A and Helicon, in Webb's notebook, Volume 2.

The vast majority of Birt's and Webb's suspects were of this nature – minutiae of the Moon. However Webb pointed out, on the south and south-east slope of Copernicus, a region of greater interest, 'thickly studded with very minute craters, not represented by Beer and Mädler'. His conclusion shows both the nature and scope of his expectations – widespread eruptions might still take place on the Moon:

> The omission is chiefly remarkable, as [these craters] form a continuation of the extraordinary assemblage of similar foci of eruption lying between Copernicus and Eratosthenes. This latter wonderful district, it may be

> observed by the way, has probably assumed its present honeycombed aspect during the present century, as it is hardly conceivable that in such a situation it should have escaped the persevering scrutiny of Schröter, and been left for the eye of Gruithuisen in 1815.[16]

To settle the matter once and for all was not something to be achieved piecemeal or easily; increasingly, it was obvious that it required nothing less than a complete remapping of the lunar surface. As a basis for discussing the vexed question of change, Beer and Mädler's chart was no less unsatisfactory than Schröter's fragments had been (and Webb took as much delight in pointing out the limits of Beer and Mädler's survey as Beer and Mädler had taken in excoriating Schröter's deficiencies of draftsmanship and inaccuracies). Recalling his 1859 paper on Cichus and Messier, and regretting that he had not, in the meantime, entered any further into 'this curious subject', Webb gave various excuses – not least the perpetual British affliction of cloudy skies. But an equal hindrance had been the inadequacy for the purpose of Beer and Mädler's map:

> I have uniformly referred, for the purpose of comparison, to the great map of Beer and Mädler. In doing this, I have been obliged to notice, with concern, occasional indications of a want of that high accuracy which might have been expected from the general character of the work. I have never entertained a doubt as to the correctness of their triangulation, or of the laying down of ... primary points ... but it has appeared to me that an equal amount of dependence cannot always be placed upon the subsequent filling in ... of lesser details. This is much to be regretted ... It seems generally admitted that no alteration of any considerable magnitude has taken place upon the lunar surface since the date of anything which can be called accurate observation; it is only in the smallest class of craters ... that we can reasonably look for traces of continued activity; and it is precisely there that we become sensible of a deficiency in the work of the illustrious German astronomers.[17]

Webb recalled Beer and Mädler's assertion (later retracted by Mädler) that no craters were discernible in Sinus Aestuum. It would never have been made, he declared, if only they had consulted Lohrmann's chart. In this and many other cases, Webb believed that the accuracy of their map suffered 'from their unwillingness to be under obligation to [others]'.[18]

It seems, from our remote vantage point, that Webb protested too much. And yet the tone of his comments serves to underscore how great had been the aura of infallibility which had so long surrounded the work of Beer and Mädler. For thirty years their map had been the last word. Now, active lunar observers such as Webb and Birt were discovering what should have been apparent all along – that Beer and Mädler had not mapped, once and for all, the lunar surface, even to the extent allowed by their limited means. Though theirs was the highest standard available, it was still a human standard, attained in a comparatively short time; it demonstrated all the failings and imprecisions of their methods of visual observation and draughtsmanship, and attested to the small aperture of their telescope. 'It is the unfortunate want of confidence' in Beer and Mädler, Webb lamented, 'which throws some doubt upon the evidence of change ... and renders an appeal to the future still necessary to render it conclusive.'[19]

To define the limits of previous work was, however, an essential part of preparing the ground work for the future, which seemed bullish – and British. The BAAS responded favourably to Birt's proposal by appointing a new 'Committee for Mapping the Surface of the Moon', with Birt himself serving as Secretary. The other members included Phillips, Sir John Herschel, Warren De La Rue, Lord Rosse and Webb. Birt's plan of action called for a collaborative and coordinated venture on a very ambitious scale, the goal being nothing less than to produce a detailed lunar map on a scale of 100 inches to the diameter of the Moon – immense compared with the 37.5-inch diameter of Beer and Mädler's map – to be achieved by dividing the Moon's entire visible surface into the four quadrants (I–IV) of Beer and Mädler; each quadrant in turn to be subdivided into sixteen 'grand divisions', designated by capital letters from A to Q; and each 'grand division' further subdivided into twenty-five areas of 5° square (denoted by Greek letters α to ω; the last space being left blank). Any object within a specified region was to be distinguished by a number given together with the quadrant and symbol for the small 5-degree area. Thus, for instance, the designation IAσ40 would indicate the fortieth object in area IAσ.

Any collaborative effort takes time to plan and coordinate. One recalls the 'Celestial Police' who attempted to organise at Schröter's observatory in 1800 for the purpose of prospecting trans-martian space for small planets. Birt's project was conceived on a grand scale, so it would take much time to complete, especially since it involved the efforts of so many observers. Using photographs provided by De La Rue and Lewis Rutherfurd, Birt prepared outline maps of three 5-degree zones, located between 0° and 6° W and 0° and 10° S. He had these outline maps printed and distributed. As a further incentive for participation he offered on loan a 7.3-inch refractor from Edward Crossley (a wealthy amateur astronomer in Halifax), which was reserved exclusively for the work of the Moon Committee. Birt set it up at Cynthia Villa – his private observatory at Walthamstow, north-east of London.

Thus began the Moon Committee's great project to scrutinise 'lunar objects suspected of change'. It was a carefully planned enterprise, promoting the establishment of a basis of observation of the physical aspect of the Moon which, in true Baconian spirit, was to be 'laid broad and deep', its 'superstructure ... characterised by accuracy and precision', so that the results arrived at might be 'beyond dispute and fully capable of testing any question that may arise as to the state of the Moon's surface'.[20]

Under Birt's enthusiastic leadership and watchful eye, the Moon Committee set out on its ambitious quest for the Grail of incontrovertible lunar change. In large part – during the remainder of the century, into the twentieth and even, so far, into our own twenty-first century – the quest for lunar transient phenomena has remained the leading preoccupation, not to say obsession, of many amateurs who have concerned themselves seriously with the observation of the Moon. Inasmuch as this is so, they have worked in the long shadow cast by the Revd T.W. Webb.

Notes and references

1 M. Rudwick, *The Great Devonian Controversy* (University of Chicago Press, 1985), p. 40.
2 Ibid. p. 37.
3 Charles Darwin, *The Autobiography of Charles Darwin and Selected Letters*, Francis Darwin (ed.), (1892; Dover reprint, 1958), p. 26.
4 T.W. Webb, *Celestial Objects for Common Telescopes* (1917; Dover reprint, 1962), vol. 1, p. 79.
5 Ibid., p. 101.
6 Ibid., p. 81.
7 T.W. Webb, 'Notice of traces of eruptive action in the Moon', *Monthly Notices of the Royal Astronomical Society*, **19** (1859), 234–5.
8 Ibid., 234.
9 Richard A. Proctor, *The Moon: Her Motions, Aspect, Scenery, and Physical Condition*, second edition (London, 1878), pp. 183–4.
10 Webb (n. 4), p. 79.
11 John Phillips, 'Suggestions for the Attainment of a Systematic Representation of the Physical Aspect of the Moon' (20 March 1862), *Proceedings of the Royal Society of London*, **12** (1863), 31.
12 Edward Sabine to John Philips, 1 March 1862, quoted in Roger Hutchins, 'John Phillips, "Geologist–Astronomer", and the Origins of the Oxford University Observatory, 1853–1875,' in Peter Denley (ed.), *History of Universities* (Oxford University Press, 1994), p. 211.
13 W.R. Birt, 'On methods of detecting changes on the Moon's surface', *Report of the BAAS*, **34** (1834), 4.
14 Ibid.

15 T.W. Webb, 'On certain suspected changes in the lunar surface', *Monthly Notices of the Royal Astronomical Society*, **24** (1864), 201–6.
16 Ibid., 204.
17 Ibid., 201–2.
18 Ibid., 202.
19 Ibid.
20 W.R. Birt, 'Report of the Lunar Committee for Mapping the Surface of the Moon', *Report of the BAAS*, **38** (1868), 1–2.

Chapter 11

Webb among the planets

Richard M. Baum

What unexplored wonders may lie in still remoter space?

In the quarter of a century or so that elapsed between the appearance of the first edition of *Celestial Objects for Common Telescopes* and the death of its author in 1885, a 'new astronomy', to use S.P. Langley's (1834–1906) evocative phrase, had grown up alongside the old.[1] Agnes Clerke (1842–1907) observed:

> One effect of its advent [was] to render the science ... more popular in its nature, because the kind of knowledge it now chiefly tends to accumulate is more easily intelligible – less remote from ordinary experience – than that evolved by the aid of the calculus from materials collected by the use of the transit-instrument and chronograph.[2]

Webb was central to this movement. Through his unpretentious style, his judicious choice of language, his avoidance of exaggeration in matters of precision, and above all, his constraint of the inferences to be deduced from isolated facts – characteristics that owed much to the influence of William Rutter Dawes[3] (1799–1868), to whom he was greatly attached – Webb transformed the novice into observer, and almost certainly influenced the way amateurs reported their observations.

Plainly, the prospect of metropolitan acclaim beckoned. But that was not Webb's world; like another cleric, the celebrated naturalist and diarist Gilbert White (1720–1793), he remained in his beloved parish, in the serenity of his own garden, to create that rare achievement, a compendium of enduring interest: *Celestial Objects for Common Telescopes*.

In spite of the deserved popularity of the book, little attention hitherto has been given to its author's development as a planetary observer, or indeed to how the book itself came into existence. Within a limited compass it is not possible to examine such matters in depth. Accordingly, a sense only of this process is sought through extracts from published writings, and unpublished notebooks and diaries. Together these provide insights illustrative of how Webb the novice astronomer – incommoded by fantasies of 'mountains and cavities on Venus' as a youth – became, in maturity, an observant scientist, assured author, and the godfather of modern amateur astronomy, counting among his friends and acquaintances such luminaries as William Rutter Dawes, Sir John Herschel (1792–1871), John Phillips (1800–1874), Lord Rosse (1800–1867), and Warren De La Rue (1815–1889).

Webb's astronomy has of course been superseded by advances in technology and is now read for its historical value. In part it is sourced in the work of the Hanoverian astronomer Johann Hieronymous Schröter (1745–1816), who laid the foundations of comparative planetology.[4] Its suppositions, therefore, though not without interest to the historian, long ago passed through the Ivory Gate. Yet we must not overlook what they represent; a vignette of the Solar System as it was perceived by the people of the time. Naturally, many of the ideas now seem naive, and even absurd. The past however, is not to be judged in terms of the present.

From the age of ten Webb kept a series of notebooks which began as exercises set by his father, and cover the years from 1817 to 1861.[5] Initially they focused on spiders and the insect world, but include a veritable treasury of solar and lunar haloes, parhelia, coronae, luminous arches, and other optical effects observed in the Earth's atmosphere. Astronomy soon crept in: on 5 January 1818 a meteor was noted; on 17 January a lunar mountain attracted his interest; and on 10 July and 11 August, Jupiter was observed. On 2 February 1819 Webb looked at M42, the Great Nebula in Orion, possibly for the first time:

> I made an observation upon Orion and saw in his sword a remarkable nebula. It had the appearance of a little white cloud, and two parts of it were whiter than the rest, which appeared like stars seen through it. The following night I observed it again. It is visible with the naked eye.[6]

In later years these observations, which signalled his astronomical debut, were transferred to other notebooks – five of which are now housed in the library of the Royal Astronomical Society (see Chapter 9).

Webb's extant personal diaries cover the period 1826–40.[7] They contain many observations of the planets – indeed, more than of the Moon and stars – but few observations of the Sun. He is a laconic diarist: 2 June 1826, 'J[upiter] out for a moment'; 31 October 1827, 'Had an observation on Saturn. Saw the body & ring distinctly'; 22 September 1831,

> Went to speak to Jones, & had a long talk w. him. came back in a hurry to take M[other] out. we went by Geer Copp to Llandinabo – just as we had come in, Dr. P. came, & stayed some time, would not stay dinner. Moon & Jupiter at night. Very fine day. Conveyed observations of Jupiter into a separate journal.

The entries may be economical, but they have the merit of precision and spontaneity.

Webb logged his first observation of Mercury, the innermost planet, on 22 May 1823.[8] In *Celestial Objects* he alludes to the difficulties of obtaining a satisfactory view of the planet: 'Though at times readily visible to the naked eye,' he says, it is 'but seldom seen from its nearness to the Sun.'[9] Imagine his delight on the evening of 3 April 1826 when he looked for and found Mercury – an experience that resonates in what he later wrote about the planet, which 'was sometimes missed, as well as once noticed extremely faint, even by the experienced Schröter'.[10] The latter was more successful than Webb, who could only remark: 'Ordinary observers will not see much.'[11] Even so, he was optimistic, and urged those of enterprise to attack the problem.

Like so many observers before and since, Webb experienced uncertainty and frustration in his transactions with Venus, little realising that all we can see from Earth is the atmosphere of the planet – specifically the bright upper surface of a visually opaque stratum of pale yellow cloud. His record is typical. On 7 November 1826, 'Obs. Venus to great advantage – fancied spots'; on 11 Octoner 1828, 'Obs. Venus – a beautiful dichotomy & thought I saw spots'; and on 21 January 1830, 'At 1h pm succeeded in seeing Venus with the naked eye and with the telescope. A beautiful object.'

Nevertheless, Webb can be credited with one astute insight. Discussing observations of the planet by Francesco De Vico (1805–1848) and his assistants, at Rome during 1839–41, he took note that the most successful in detecting markings were 'those who had most difficulty in catching very minute companions of large stars'. The

67

Observations upon Venus.

1844. Nov. 27. $7^h 45^m$ A.M. Air rather unsteady. 5f A. 112. Versed sine of the disc about 0·73. The planet seemed to have a faint freckled or mottled appearance. $8^h 15^m$ The same, with 112 & 144, but not confirmed with 250: probably too much haze and smoke.

1846. Jan. 5. 4^h P.M. Air unsteady & fluttering. 5f A. 144. Versed sine about ·5
Venus a very beautiful object. Had the atmosphere been more favourable, I fully believe freckled spots might have been seen. The cusp of the S. horn seemed to have something like a shadow upon it, not unlike that in Fig. 59. of the Aphroditogr. Fragmente, dividing it from the rest of the disc; but on account of the state of the air this was very uncertain. $4^h 45^m$ I could not trace the shadow at the cusp: but the freckled appearance of the disc was unquestionable, & it seemed as though it would have exhibited a beautiful picture, with dark spots like the moon, though less contrasted in proportion, could only a momentary steadiness have been obtained in our own atmosphere. – A freckled appearance had been previously noticed on the evening of Jan. 1.

1847. July 1. $6^h 20^m$ P.M. Air very clear. 5f A. 144. The terminator appeared a little more shadowy towards the S. horn, as in Fig. 1. This appearance was very faint, but pretty certain; & was confirmed with 112, which also shewed it again at $6^h 40^m$. It probably entered a considerable way into the disc, extending parallel to the limb; but so indistinctly & uncertainly that I have not ventured to represent it. Seen again with 112 at $7^h 5^m$. – I have seldom seen the planet better defined.

1847. July 3. 6^h P.M. 5f A. Air very clear, but vision rendered uncertain & wavering by smoke from intervening chimneys. As soon as 80 was brought to focus, some

Observations upon Venus, Webb's notebook, Volume 5, p. 67.

explanation seemed obvious enough to Webb, who in a lengthy footnote asserted: 'A very sensitive eye, which would detect the spots more readily, would be more easily overpowered by the light of a brilliant star, so as to miss a very minute one in its neighbourhood.'[12]

Here then is a rationale of why generations of astronomers failed to agree on what they saw on the apparent surface of Venus: the detection of faint shadings on the brilliant cloud surface of the planet depends more on sensitivity of vision than visual acuity, implying that some observers may be more sensitive to seeing differences in albedo and colours, while others are better able to resolve lines and edges. Percival Lowell (1855–1916) came to the same conclusion around the turn of the twentieth century, as did the distinguished French observer Audouin Dollfus half a century later.[13]

All this contrasts sharply with what Webb confided to his diary on 20 October 1826: 'Very dull day & very rainy evening, but held up at night – dream of mountains and cavities on Venus and her dark side visible, and two satellites, blue-violet.' Assuredly, he was writing with his mind on the subject; but was this really an instance of 'the images of memory flowing in on the impulses of immediate perception', to use Samuel Taylor Coleridge's evocative phrase? Webb did not first see the faint illumination of the dark side, the so-called 'ashen light' (which is still a controversial phenomenon) until the evening of 31 January 1878.[14] So this was not the Venus that shone quietly down on Gloucester. It was something of which he had read; of truth refracted by a prism of analogy and inference; a vivid imaginative rendering of what the earlier observers such as Burratini, the elder Cassini, Derham, Fontana, La Hire, Riccioli and Schröter believed they had seen. Of rugged cordilleras, and towering mountains, shimmering eerily in the light of an apochryphal Moon adrift in an ashen sky. But this is curious, for Venus travels alone. So, what did Webb have in mind when he set down in his diary, ' ... and two satellites, blue-violet'? This is a planet of the books, of other men's visions and thoughts; yet it is also a transcript of immediate perception like those vigilant records amongst which it is to be found.

At this point it is well to remark that spaceprobes, beginning with Mariner 2 in 1962, have not brought to light any substantial object in the vicinity of the planet. Nevertheless, during the nineteenth century the satellite of Venus had its supporters, even though it was last 'seen' in 1768. Before then, beginning in 1645, it had been observed on thirty-three occasions by fifteen astronomers. What these observers saw was finally explained by Paul Stroobant (1868–1936) at the Royal Observatory of Belgium at Uccle in 1887.[15] After a rigorous analysis of all the reports he concluded that in each case the observer had been mistaken, confusing nearby faint stars as evidence of satellitic companions of the planet. There was no satellite – and there never had been. Whether discrete objects a few metres in diameter existed was quite another matter, though it was unlikely. To all intents and purposes, Venus is moonless.

This is what the sixth (1917) edition of Webb's *Celestial Objects* tells us. The text (edited by Espin) has been sanitised; but earlier editions are more forthcoming: 'This is an astronomical enigma.' According to Webb:

> It is not easy to set aside the evidence of its occasional appearance; but it is no more difficult to understand why, if it exists, it is so seldom visible, for the diameter ascribed to it is about ¼ that of Venus ... Humboldt classes it with the 'myths of an uncritical age'. Smyth inclines to an opposite opinion. Hind considers it 'a question of great interest,' and says it 'must remain open for future decision'.[16]

Humboldt, of course, was correct. But there is allusion in what Webb wrote; a sense of anticipation, a feeling no more, of something yet to be discovered. We are not disappointed. In the pages of his notebooks we find the reason: a four-page account, dated 10 October 1832, of what is intriguingly called 'Old Observation on the Satellite of Venus, inserted here in order that it may be preserved'. The account, written in a neat hand, describes an experience Webb had had some nine years earlier. The following extract supplies the pertinent detail:

> It happened to me while a youth, to observe an appearance so similar to the supposed Satellite of Venus, that I now greatly regret that I did not record it at the time. Whatever may be its value, I am at any rate desirous that it should not be lost, and sorry that it should have been so long entrusted to memory, and therefore I have inserted in this place such particulars as I can recollect concerning it. I believe that at the time when it occurred I had never read any account of the satellite of ♀ [Venus] though I cannot be quite certain as to this point. I have the strongest possible impression of it, which is corroborated by the fact that it did not excite any very extraordinary interest in my mind, & that I made no memorandum of it at the time in a book in which I recorded remarkable appearances, as I certainly should have done had I been aware of the previous observations of some astronomer as to this point. The only record I have of it is in a pocket book of the year (1823), where in the margin, opposite May 22, stands this mark which I know refers to the observation in question. Until that evening, I had never seen Mercury, & finding from White's Ephemerides that he was then in a favourable situation & the evening being clear, I went into the garden of our house at Gloucester, with a little achromatic by Bate, 14 inches

> focal length, 1.3 inch aperture, & 22 inches when drawn out with a terrestrial eye-piece of 4 glasses magnifying of course very slightly – I think thirteen or fourteen times, but capable of exhibiting to my practised eye, the nebula in the Triangle [M33]. The defect of my sight, as I did not then wear a concave glass, prevented my seeing the planet without the telescope, but with it I soon found him, low in the W. horizon, & upon turning the glass to ♀ [Venus], then high in S. W. I observed a star exactly like ☿ [Mercury], or like Venus in miniature, preceding or south preceding the planet at a short distance, perhaps 20' or 30'. It was much smaller than ♀ – possibly ⅓ or ¼ of its diameter; but with so inadequate a power the disc of ♀ could not of course be well seen. I have a very distinct recollection of its great resemblance to ☿.[17]

Although he searched on succeeding evenings, Webb never again saw the mystery object; but years later, when he finally published an account of the observation, he obtained a degree of clarification:

> Through the kindness of Mr. Lynn I have been enabled to ascertain that the star ε Geminorum was not far from the planet [Venus] on that day, only about 30.5' further S., which would agree very fairly in that direction, but lying 6.5m. more to the E.[18]

Was this the explanation? Most probably. Nevertheless, a vestige of doubt remained:

> Independently of this discrepancy – a serious one, for I have no doubt of the *p* [preceding] or *sp* [south preceding] position of the satellite, not only clearly remembered but shown in the little diagram – it does not seem probable that a star of 3–4 mag. should have been so conspicuous in such an instrument in the twilight. I have no note of the hour, but as Mercury had not sunk into the smoke of the town [Gloucester] in the W. horizon, it must have been comparatively early, and at that time of year the twilight is strong. It may be too hazardous under all the circumstances to include this with the other observations of the pseudo-satellite, but there seems no reason why it should pass into entire oblivion.[19]

Venus as a comet? No, this is not a reference to the ideas of the Russo-American psychiatrist Immanuel Velikovsky, though it might well be. Rather, it concerns another deletion from later editions of *Celestial Objects*: the remarkable story, told by J.H. Mädler, relates how, in the field of his 4-inch refractor, he saw brushes of light diverging from the cusps of Venus in the direction of the Sun, giving the planet the appearance of a comet.[20]

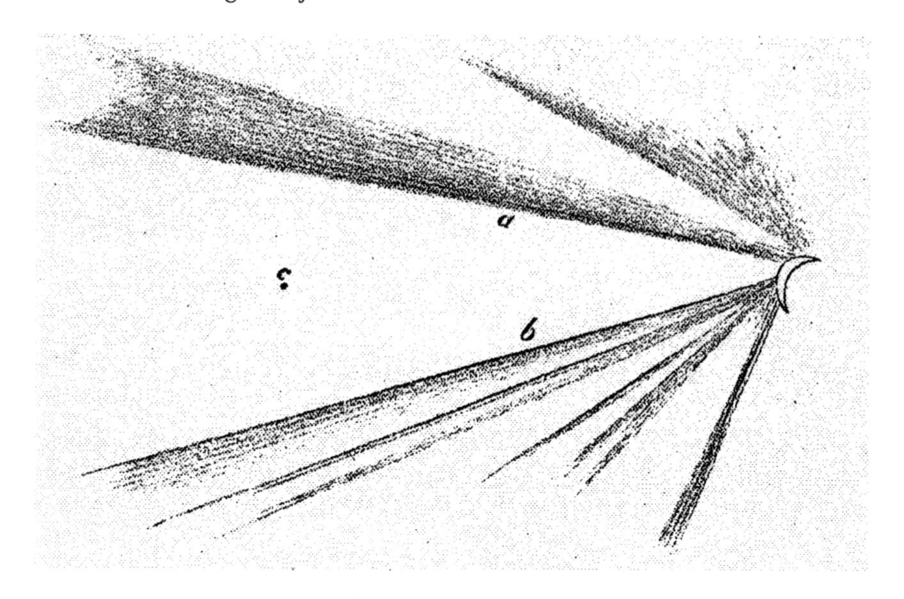

Venus with 'light brushes', observed by J.H. Mädler with a 4-inch refractor on the evening of 7 April 1833. (From W. Beer and J.H. Mädler, *Bertrage zur Physischen Kenntrtiss der Kummlischen Korper im Sonnensysteme* (Weimar, 1841).)

Webb supplies no detail, but Mädler cites the observation, made 7 April 1833, in his *Beitrage* (1841). Venus was an evening object and well placed for observation. No explanation is attempted, but he was disinclined to regard it as an optical illusion. Even so, illusion is the first thought that comes to mind. One leading American astronomer of recent times went so far as to ascribe it to a greasy thumbprint on the telescope objective. Did Webb have a similar thought? We cannot say. He is ambivalent, and almost conciliatory, but comments: 'It is an instructive instance of the oversights which may be incidental even to great philosophers, that it never seems to have occurred to him to try another telescope.'[21] It seems curious, therefore, that Webb should choose to include the observation at all, given its dubious character. Again his early records are revelatory. The first book of his 'Observations upon Nature' contains the following entry, written when he was thirteen:

> 1820. Feb.18. At about 7h P.M. I was called upon to witness a singular appearance. A tail had been seen to extend downward from the planet Venus which was then at a considerable length and shining very brightly. This tail was the length of the Moon's apparent diameter, but was not visible when I went out. However, soon afterward I again saw it distinctly and a ray of light not so considerable indeed as I was informed at first,

> but very bright and plain, at times appearing and disappearing on another side of the planet. Sometimes it pointed upwards and sometimes at the side. Being viewed through a telescope, Venus appeared elongated perpendicularly when the tail pointed upwards or downwards, so much that at one time she seemed three or four times as long as broad and when the tail was sideways she seemed to project in that direction, and at one time was magnified nearly as large again as her common size. Looking at her about an hour afterwards I again saw the same appearance which seems to have been produced by some vapour in the air. This I observed not infrequently afterwards.

Webb, then, was also a witness to the masquerade – and not the only witness. We can refer to the diary of John Gadbury (1627–1704), kept between November 1668 and December 1689, which he declared was 'design'd for the services of philosophers, physicians, astrologers and all other faithful observers of the various wonders that are to be found in God's creation'.[22] Though little known, this diary is a valuable record of late seventeenth-century London weather, and also a fascinating register of atmospheric optical effects observed during that period. Running a finger down its columns we discover that on 29 January and 21 February (Old Style) 1686 he recorded 'Venus like a comet'.[23] What, therefore, can be made of the greasy thumbprint hypothesis? – especially when the published correspondence of Kepler tells us: 'My countrymen maintained that they have seen at times Venus as a comet' – a comet being an omen of threat and disaster.[24] It seems unlikely that Webb was aware of the effects described by Gadbury and Kepler, and it is also strange that in reporting what Mädler saw he should omit his own observation. As for the brushes of light: we must seek an answer in the physics of the atmosphere – but so far there is no satisfactory explanation of the mechanism involved.

What does this tell us about Webb and the planets? It is an affirmation of his inherent skill in observation and his attention to detail – further evidence of which is also to be found in the rationale underpinning his renditions of celestial objects. Like Dawes he was a good draughtsman, but his depictions of the planets were made on a smaller scale than commonly adopted. Webb reasoned, quite logically, that if the observer draws on a large scale there is not only the temptation to insert rather more detail than is actually observed, but a tendency to misrepresent appearances by giving emphasis to features otherwise indefinitely seen.[25]

The result of his philosophy is an accurate portrayal of what the owner of a modest telescope can reasonably expect to see at similar resolution – a fact exemplified by a panel of thirty drawings of Mars in one of his notebooks. Dusky markings such as the Margaritifer Sinus,

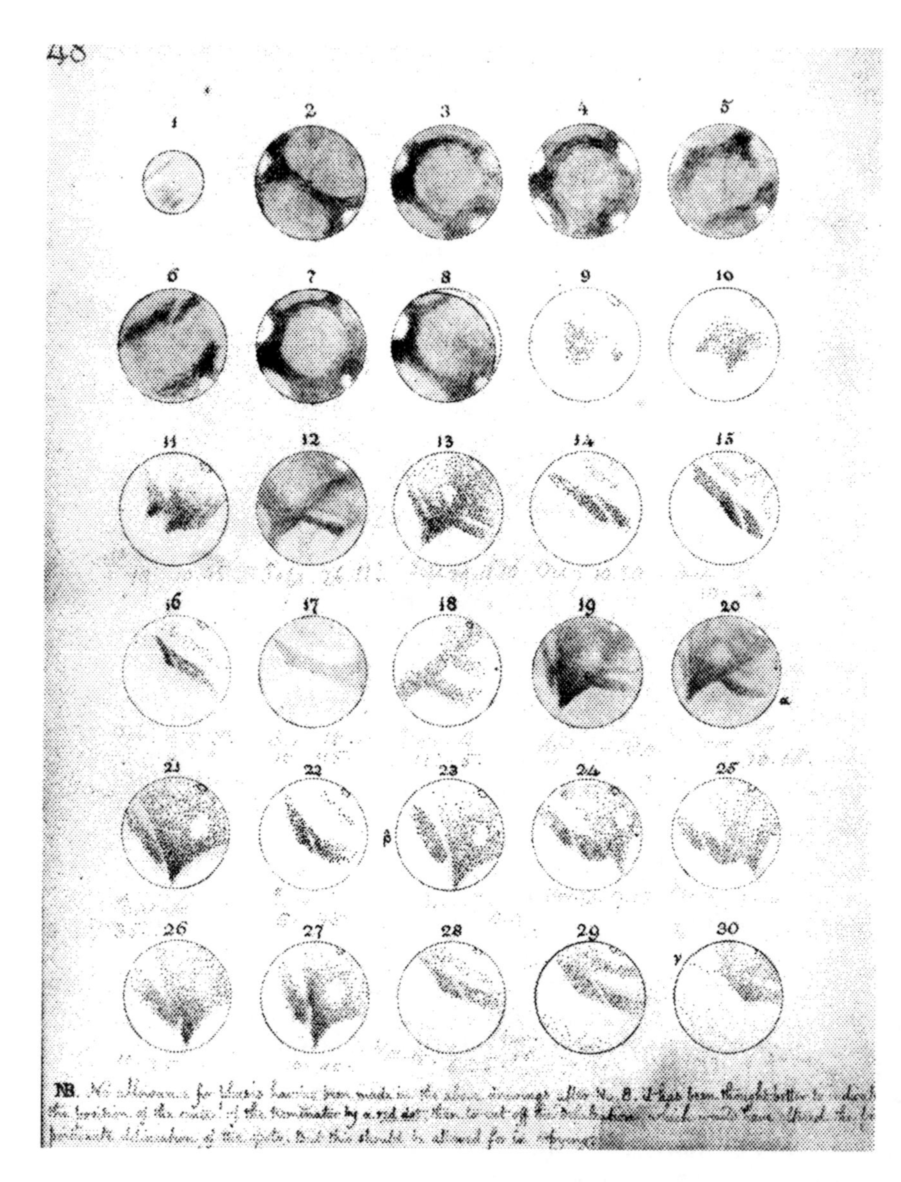

Thirty drawings of Mars, in Webb's notebook, Volume 5, p. 48.

Mare Cimmerium, Syrtis Major, Sinus Sabaeus, and the bright spots Hellas and Noachis and so on, are exquisitely depicted, all in near perfect proportion, precisely as they appear to the user of a small telescope. The south polar cap is conspicuous in the drawings – its regression clearly marked, from a large oval spot to a tiny almost circular speck. On 4 October 1830 he noted: 'At night saw the snow at the pole of Mars.'

Here the 'malignant aspect' of the planet is transformed 'into that of a miniature Earth', he later wrote, 'which we might, without much extravagance, imagine to be habitable by man'[26] – a tantalising insight which is perhaps best expressed in what he had to say in 1880 about two maps of the planet drawn up from observations in 1877 by the English artist–astronomer N.E. Green (1823–1899) at Madeira, and G.V. Schiaparelli (1835–1910) at Milan. The former saw no trace of the *canali* so characteristic of the latter's work – the infamous 'canals':

> Green, an accomplished master of form and colour, has given a portraiture, the resemblance of which as a whole, commends itself to every eye familiar with the original. The Italian professor, on the other hand, inconvenienced by colour-blindness, but of micrometric vision, commenced by actual measurement of sixty-two fundamental points, and carrying on his work with most commendable pertinacity, has plotted a sharply-outlined chart, which whatever may be its fidelity, no one would at first imagine to be intended as a representation of Mars.[27]

In other words, Green had produced a 'picture' of the planet, whereas Schiaparelli had drawn a 'plan'. To Webb this dichotomy represented the end points of a spectrum, and in selecting a map of Mars for the fourth (1881) edition of *Celestial Objects* he chose that drawn in 1879–80 by Charles Burton (1846–1882) and J.L.E. Dreyer (1852–1926), because, he stated:

> Most ... features are represented in a design ... which, while free from the conventional hardness that renders the elaborate designs of Schi[aparelli] so perplexing to the student, resembles in all essential points the beautiful delineation of Green, combining, however, many indications of the curious network of narrow streaks, which the last-named observer considered too uncertain or transitory to be introduced without further confirmation, and the nature and permanence of which will be an interesting object for future investigation by the possessors of adequate optical means.[28]

Here Webb neither rejects nor accepts the *canali*, but as a compromise almost hints at pluralism – the possibility of life elsewhere in the

Universe. Was this influential in his choice of Burton's map, as Michael Crowe suggests? Or was it in keeping with the elder Herschel, who was swayed by presumed analogies with Earth?

In 1870 Webb wrote:

> It is fortunate that Jupiter, the nearest at once and the largest of that external group [of planets], presents a disc so broad and so luminous as to invite examination even with telescopes of moderate size.[29]

Certainly his results for the years 1869, 1870 and 1871 conform to this expectation. Using his $9\frac{1}{3}$-inch With–Berthon reflector, his first series of observations began on 15 October 1869 and continued until 11 March 1870, during which period Jupiter 'was examined, though sometimes to little purpose' on forty nights.[30] In the second series, which commenced on 15 November 1870 and ended on 15 April 1871, the planet was scrutinised on forty-nine nights.[31] 'In order to render descriptions more ready and intelligible', he reported:

> I have thought it allowable to assign, for my own purpose, and for the present season, names to the different features of the planet. I have called the brighter stripes, *zones*, the darker, *belts*. The central portion, which has recently exhibited such singular details, I have termed the *equatorial zone*; its two dark borders, the *north* and *south torrid belts*; the two bright regions on either side of these, the *north* and *south temperate zones*, the latter being subdivided by a dusky stripe, which may be called the south subtorrid belt; beyond these two zones respectively lie the *north* and *south temperate belts*; and on the further side of these, the two *polar regions*.[32]

Thomas Hockey, of the University of Northern Iowa, has recently conducted research into pre-photographic studies of Jupiter, and in his book *Galileo's Planet*[33] he notes how Webb complained that so few astronomers could draw really well, besides considering at length other aspects of his jovian studies. But Webb was as much concerned with the physical condition of the gas giants as with morphology. In the *Popular Science Review* of 1870 he stressed:

> An important distinction, has been repeatedly pointed out, between the group of interior and exterior planets, as referred to the wide interval now known to be occupied by a multitude of minuter bodies. Either group, as far as observation extends, or fair analogy will carry us, has a character peculiarly its own; the outer being distinguished from the inner by inferior comparative density, but superior magnitude, velocity of rotation, and attendance of satellites. These remarkable differences, though increasing the interest of such researches as may be permitted to us by the Great Ruler of the universe, add materially to their difficulty; and we find it impossible

to carry on to remote planets the analogies which have apparently served us so efficiently in the case of our closer neighbour Mars.[34]

Writing in the same periodical the following year he extends the discussion to considerations about the physics of the jovian atmosphere in the course of which he remarks:

It is not improbable that a considerable portion of the heat of Jupiter may be of an unborrowed character. This idea, which has been advocated on other grounds, seems favoured by another circumstance. Were his temperature due merely to solar radiation, the currents ascending from the hotter equatoreal regions would be observed to deviate to some extent in an oblique direction, their lateral diffusion being unrepressed by equivalent expansion in remoter latitudes; the phenomena, however, afford but very equivocal instances of any such tendency, the oblique arrangements which are occasionally visible being too irregular in aspect or inconsistent in direction to be referred with safety to this origin.[35]

The core of this speculation was again raised during the twentieth century.

Webb includes much about Saturn and its magnificent ring system in *Celestial Objects*. 'Fortunately for the student,' he says, 'a common telescope will exhibit some part of the wonders of this superb planet.'[36] Yet his extensive chapter is more informative of what others have seen than a review of his own studies. While it displays his erudition, it provides no significant insight into his personal thoughts or recollections of the planet, although in his observations he commented on a bi-tonal aspect of the shadow of Ring A – sometimes black on its inner edge, and grey at the outer edge.

Writing in 1863, he says Herschel's planet was seen with the naked eye as a 'very minute but sharp and steady point.' At the time, Uranus was in the constellation Taurus, somewhat below the line from β Tauri to Aldebaran. In the same article he noted that a large telescope was necessary to show the polar flattening, adding that the planet seemed more luminous than expected, considering its great distance from the Sun.[37] He enlarged somewhat on this in *Celestial Objects*:

Uranus being visible in clear weather to the naked eye, will be easily caught up in the finder by the help of the almanac, and will be large and planetary-looking in the telescope; its disc indeed subtends less than 4 inches, but I never found the light of $3\frac{7}{10}$ in. sufficient to define it perfectly; $5\frac{1}{2}$ in. dealt far better with it: with Lawson's 7-in. object-glass, bequeathed to the Greenwich Naval School, I have seen it beautifully, as a little Moon: no one has made out much more.

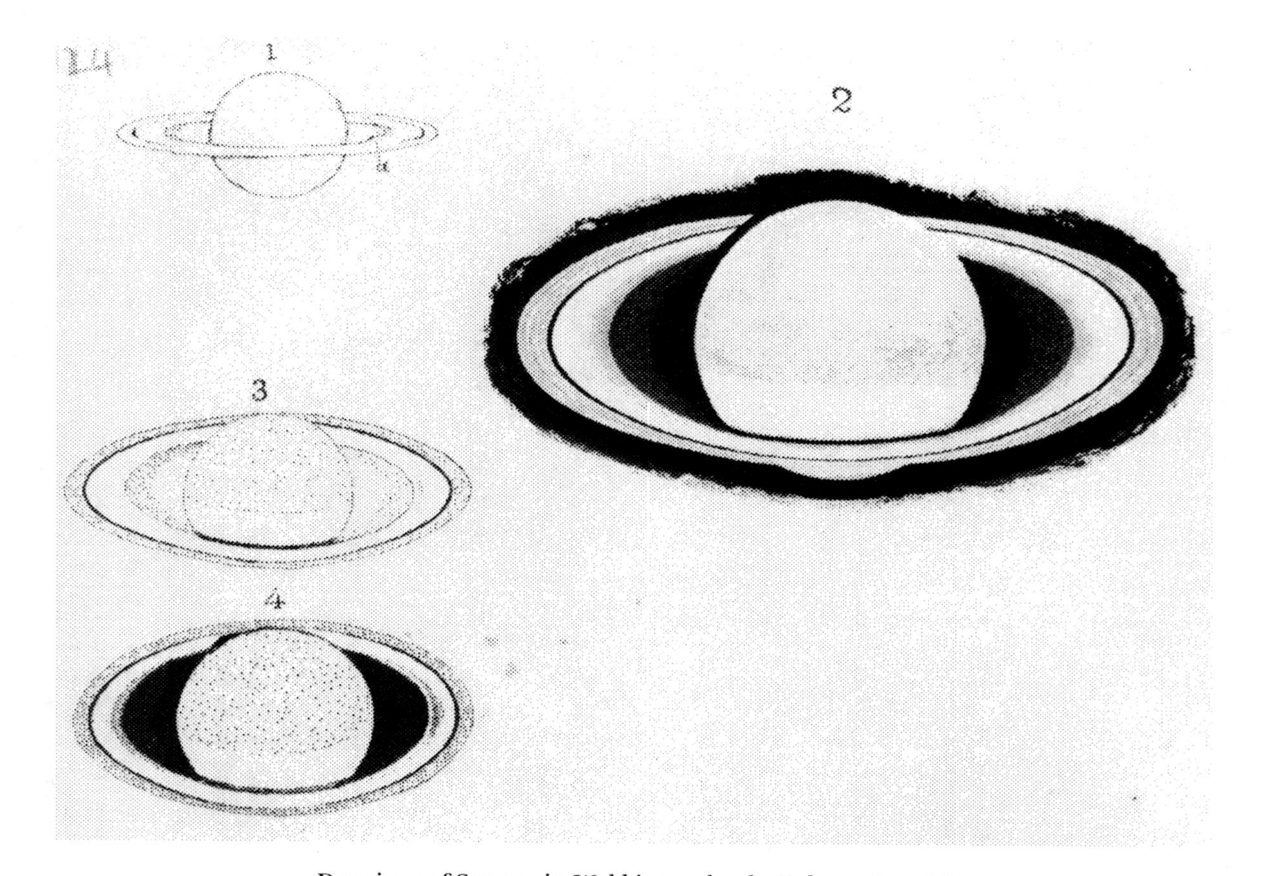

Drawings of Saturn, in Webb's notebook, Volume 5, p. 124.

With regard to the satellites then known, he says: 'My less acute eye has just caught one with $9\frac{1}{3}$in. speculum.'[38]

His comment about Neptune is just as laconic: it 'will hardly repay the search'; and on the occasions when he *did* see it he found it 'dull and ill-defined', although he qualifies this by interjecting a speculative hint of the then burgeoning transneptunian hypothesis:

> But who can say how grand a spectacle this inconspicuous globe might present on a nearer approach? or what mysteries might be developed in a spectrum much resembling, according to Se., [Secchi] that of Uranus? or that he is the most distant planet that obeys our central Sun? or what unexplored wonders may lie in still remoter space? ... The advance of optical power may be expected either to open up fresh marvels, or prove to us that, as far as the dominion of our own Sun is concerned, we have reached the boundary of our knowledge.[39]

In conjectural mood, as in his observations, Webb showed himself to be gifted and thoughtful: his observations were illustrated by admirable drawings and distinguished by scrupulous accuracy. It is these characteristics that shine through the pages of *Celestial Objects for Common Telescopes* – transforming it from a mundane compendium into an inspirational classic of astronomical literature.

Notes and references

1 S.P. Langley, *The New Astronomy* (Ticknor, Boston, 1888).
2 Agnes M. Clerke, *A Popular History of Astronomy during the Nineteenth Century* (Adam and Charles Black, Edinburgh, 1885).
3 William Rutter Dawes – the 'eagle-eyed' – discovered Saturn's crêpe ring, independently of the American William C. Bond, in 1850; see R.A. Marriott, 'Dawes, Lassell and Saturn's dusky ring', *Journal of the British Astronomical Association*, **96** (1986), 270–7. Dawes' observations of Mars were used by R.A. Proctor to construct the first reliable map of the planet; see R.J. McKim and R.A. Marriott, 'Dawes' observations of Mars, 1864–65', *Journal of the British Astronomical Association*, **98** (1988), 294–300.
4 W. Sheehan and R. Baum, 'Observation and Inference: Johann Hieronymus Schröter, 1745–1816', *Journal of the British Astronomical Association*, **105** (1995), 171–5.
5 See Chapter 1, n. 3.
6 T.W. Webb, 'Observations upon Natural Phenomena. Part I', 7 August 1817 to 30 November 1821.
7 For a physical description, see Chapter 1.
8 T.W. Webb, 'The Satellite of Venus', *Nature*, **14** (1876), 193–5.
9 T.W. Webb, *Celestial Objects for Common Telescopes*, fourth edition (1881), p. 46.

10 Ibid.
11 Ibid.
12 Ibid., p. 56.
13 G.P. Kuiper and B.M. Middlehurst, *Planets and Satellites* (University Press, Chicago, 1961), p. 552.
14 T.W. Webb, 'Astronomical Notes, Letter 13938', *English Mechanic*, **26** (1878), 627.
15 For P. Stroobant, see *Étude sur le Satellite Énigmatique de Vénus. Memoirs Couronnesde la Academie Royal des Sciences de Belgium*, **44** (1887).
16 Webb (n. 9), pp. 61–2.
17 T. W. Webb, 'Observations on the Moon, 1826–55.' The account immediately follows an observation, dated 8 October 1832, on the lunar formations Mare Crisium, a rill near Hyginus, and Hevelius.
18 Webb (n. 8).
19 Webb (n. 8).
20 Webb (n. 9), p. 62.
21 Ibid.
22 R. Baum, 'The Mädler Phenomenon', *The Strolling Astronomer*, **27** (1978), 118–9, reprinted in W.R. Corliss, *Mysterious Universe: A Handbook of Astronomical Anomalies* (The Sourcebook Project, Len Arm, Maryland, 1979), pp. 126–8.
23 Ibid.
24 C. Baumgardt, *Johnnes Kepler: Life and Letters* (London, Victor Gollancz, 1952), p. 47.
25 A.C. Ranyard, obituary of T.W. Webb. *Monthly Notices of the Royal Astronomical Society*, **46** (1886), 198–201. See also R. Baum, 'T. W. Webb', *Dictionary of Nineteenth-Century British Scientists*, 4 vols. (Thoemmes Continuum, 2004), Vol. 4, pp. 2126–9.
26 Webb (n. 9), p. 140.
27 Webb, 'The Planets of the Season: Mars', *Nature*, **21** (1880), 212–3.
28 Webb (n. 9), p. 144; also W. Sheehan, *Planets and Perception: Telescopic Views and Interpretations, 1609–1909* (University of Arizona Press, 1988).
29 T.W. Webb, 'The Planet Jupiter, 1869–1870', *Popular Science Review*, **9** (1870), 127–37.
30 Ibid.
31 T.W. Webb, 'The Planet Jupiter in 1870–71', *Popular Science Review*, **10** (1871), 276–83.
32 Webb (n. 29), p. 128.
33 T. Hockey, *Galileo's Planet: Observing Jupiter Before Photography* (Institute of Physics, Bristol, 1999).
34 Webb (n. 29), pp. 127–8.
35 Webb (n. 31), p. 281.
36 Webb (n. 9), p. 171.
37 *Intellectual Observer*, **3** (1863), 51, 124.
38 Webb (n. 9), pp. 186–7.
39 Ibid., p. 187.

Chapter 12

The comets observed by Webb

Jonathan D. Shanklin

Thou wondrous orb, that o'er the northern sky
Hold'st thy unwanted course with awful blaze!
Unlike those heavenly lamps, whose steady light
Has cheered the sons of earth from age to age,
Thou, stranger, bursting from the realms of space
In radiant glory, through the silent night,
Thy tresses streaming like the golden hair
Of Atalanta, or that beauteous maid
Pursued by Phœbus, upward shalt invite
Many a dull brow unus'd on heav'n to turn,
And many a bosom read with deep alarm.

Revd John Webb (1811), in T.W. Webb,
Celestial Objects for Common Telescopes

Webb lived in a century that was later renowned for a succession of bright comets, which perhaps went unrivalled until the late 1990s. The introductory poem by his father probably represents the Great Comet of 1811, which is ranked amongst the most impressive in history. Recently we have had spectacular objects such as comets Hale–Bopp and Hyakutake, and many a little less bright, such as Ikeya–Zhang and Machholz. Just as these comets have had an effect on today's amateur astronomers, so too did the comets of Webb's generation. Webb's life saw the transition from comets being objects of superstition to comets being objects of scientific study. When Webb was born only Halley's comet was known to return, but by the time he died there were twelve comets of known period. They had become moderately well understood members of the Solar System, with growing ideas about their nature and composition.

Webb's first detailed cometary notes[1] refer to the Great Comet discovered on 29 December 1823 by Nell de Breautè, observing at Dieppe in France (1823 Y1), with Pons and Biela, amongst others, making independent discoveries the next day. Word clearly spread rapidly, as Webb began to look for the comet on 6 January, although he did not recognise it as such: 'I had never seen a comet since I was quite a child and that I am monstrous short sighted'. He did, however, see it, noting that it was near β Herculis and had a tail a degree or two long pointing to the north-west. Modern calculations place it some 4° from β Herculis, with a tail in position angle 320°. (In many ways I had a similar experience, rising very early one spring morning in 1970 to look for comet Bennett. I never saw it, despite the fact that it was a bright object, but it was an experience that generated a desire to learn more about the night sky and engendered a life-long interest in comets.) Later in the month Webb was 'informed again that it was visible between the two bears' (Ursa Major and Ursa Minor). He noted that through the telescope 'it was of a round form, & a reddish colour. The tail was short and not very visible through a telescope'. The comment that it was of a reddish colour is interesting, as in general comets either appear a cream colour, caused by sunlight reflected from dust in the comet's coma, or a blue–green colour, caused by emissions from the (spectroscopic) Swan bands of carbon.

Given that he was just sixteen when he saw this comet, and that he had not seen a comet since he was a child, Webb's first comet was probably the Great Comet of 1819 (1819 N1). This was a widely observed evening object in the summer of that year, and would undoubtedly have made an impression on the twelve-year-old boy.

Webb's involvement with the observation of comets resumes with his notes on the comet of 1845 in 'Astronomical Observations made by Thomas William Webb'.[2] He notes that whilst his wife and servant could see the comet very clearly on 11 June, he found it a much more difficult object. He suggested that this might have been because 'my eye is less impressible by a very faint light'. Such an explanation is certainly possible, and whilst some observers find it easier to discern fine planetary detail, others find it easier to pick up the faint enhancement of nebulous objects. (This may be one reason why I find comet observing relatively straightforward, and when showing others a faint comet in the great Northumberland refractor at Cambridge University Observatory often get the response 'I can't see anything'!) Comet 1845 L1 (Great June Comet) was discovered at the beginning of June, and became widely observed over the next week or so. The elongation from the Sun decreased after the solstice, and the last sighting was on 2 July. It is a little surprising that it was not observed earlier, as it was low in the northern sky in the early morning from the end of May, and would have had a prominent tail. This may just reflect that observations of morning objects, even when they are bright, are far less frequent than for objects conveniently placed in the evening sky. A further point is that many astronomers ignore the northern horizon – particularly in summer with the long northern twilight. UK latitudes do, however, offer an opportunity at this time of year to discover a bright comet at relatively small elongations due north of the Sun.

In 1805 Jean Louis Pons discovered a comet which was found to move in a rather uncertain elliptical orbit. Wilhelm von Biela discovered a comet in 1826, and after sufficient observations had accumulated he noted that its orbit was similar to that of the comet of 1805. It was predicted to return in 1832, and duly did so. The next return was unfavourable, but it was recovered in 1845. Astonishingly, observers noted that it appeared double in mid-January. Webb observed the 'Twin Comet of Biela' several times, first recording it on 25 February. He 'found this wonderful object, & perceived its double form', but encountered a common problem for amateur astronomers, as he 'could only observe it for a few minutes, from the nearness of a roof'. He had similar problems on 2 March, complaining that the 'presence of friends, & the immediate contiguity of the roof of the cloisters of the Cathedral [Gloucester], prevented all attempts at measurement or prolonged observation of this object'. The twins returned in 1852, but in 1865 nothing at all was seen. It was thought that the comet had completely disintegrated, and in 1872, and for several returns thereafter, there were spectacular displays of meteors. This helped confirm the link between comets and meteors. The comet has never been seen since, and is known as 3D/Biela – indicating that it was the third periodic comet to have its orbit determined, and that it has since disappeared.

On 4 July 1847 F.V. Mauvais, in Paris, discovered a comet, independently discovered by George P. Bond, in the USA, on 15 July. On the same day Webb observed the comet, then lying between α and β Ursae Minoris, and was well aware that it had been discovered on 4 July – demonstrating the speed of communication across Europe. Webb continued to observe it until 9 September.

In April 1849, two comets were visible. K.G. Schweizer first sighted comet 1849 G1 from Moscow on 11 April, but Webb records observing 'the comet discovered by Mr Graham on 14 April' on 26 April. Graham made an independent discovery from Markree Observatory in Ireland. The other comet was much brighter, and Webb notes: 'M. Goujon's

Comet is a much more interesting object, perhaps just visible to the naked eye . . . With 64 [magnification] it is a beautiful pale uniform light, very little condensed towards the centre, & of a phosphorescent or electrical aspect.'

German astronomer A.C. Petersen discovered a faint telescopic comet in early May 1850. It was approaching both the Earth and the Sun, and was closest to the Earth in mid-July, but was heading south towards perihelion and was last seen at the end of the month. Webb picked it up on 7 July, and noted that 'as the sky grew darker, it was distinctly & pretty readily visible with the naked eye, as a star of the 5th magnitude'. He also commented that 'it was much brighter than the nebula in Andromeda [4th magnitude], but I fancied it more fluctuating in the character of its light'. This highlights a problem in the validation of early magnitude estimates of comets, which often refer to the nuclear magnitude rather than to the total magnitude. On the 16th he commented: 'Its light was remarkably fluctuating, producing an effect which might almost have been fancied analogous to a kind of combustion or inconceivably rapid evolution of phosphorescent but imperfectly transparent matter from a central luminous point.' Reports of rapid variation in the central coma are quite common in nineteenth-century accounts, but have not been verified by modern high-speed photometry. This is an area that amateurs could investigate, as professionals do not expect rapid variation in the light and therefore do not carry out such measurements.

Somewhat surprisingly, Gary Kronk makes no mention of any of the above observations in his monumental *Cometography*,[3] and Webb first features in his discussion of comet 1854 F1 (Great Comet). Webb's report of his observations in the RAS *Monthly Notices*[4] is slightly at variance with his own notes. In the *Monthly Notices* he says that it was first seen at Ross in the county of Hereford by 'an intelligent little boy of ten years old, who noted "a stranger in the sky"'. In Webb's 3.7-inch Tulley it 'appeared of a pale orange or fiery hue, and of an aspect which would have been deemed in former times terrible and portentous'. He also noted:

> On the first evening of my observations, as well as frequently afterwards, I thought there was, to the naked eye, aided only by a concave lens to correct near-sightedness, an appearance of blazing up or flashing, issuing from the head into the denser portion of the tail; of this Mrs Webb, whose eye is very sensitive to light, was on several occasions perfectly confident. These coruscations could not however be traced in the telescope. Mr Purchas, at Ross, observed nearly the same phenomena and on March 28 considered the nucleus as brilliant as β *Arietis* [2.6], with a tail extending 3° in the strong twilight.[4]

In his 'Astronomical Observations ... ' Webb records, on 29 March, a slightly different version:

> Mr W. H. Purchas, of Ross, having informed me by letter that a Comet with a nucleus as bright as a 2nd mag star, & tail of some 3° had been seen by him for a few minutes the night before.

Webb refers to it as the 'Comet of March, 1854 (Russian War)':

> The first glimpse, however, was very striking, from its angry yellow hue – & however we may smile at the superstitions of our forefathers, it was impossible not to be impressed by the coincidence of the sudden appearance of this drawn sword in heaven, with the Declaration of war after so long a peace – a war of which no men can foresee the ultimate result on termination.

The 'Russian War' is the Crimean War of 1854–56, during which Britain and France were allied against Russia to prevent its further western expansion, and when Florence Nightingale performed her heroic feats of nursing.

The next few comets received only brief notes. First came D'Arrest's comet (1857 D1) seen on 23 March, then 1857 F1 on 15 and 16 April. This was 5D/Brorsen, but Webb does not mention that its appearance was predicted. Several more, readily observable, comets were observed elsewhere, but are not recorded by Webb, and he next has a brief note on what is now known as 8P/Tuttle.

His notes from September 1858 describing comet 1858 L1 (Donati) are much more voluminous, and his observations are reported in the November 1858 issue of the RAS *Monthly Notices*.[5] On 12 September he noted: 'This was a beautiful object as the night drew on – with a large 3rd mag. head, & a tail of about 4°.' On 30 September he made the 'First Trial of the Great Object Glass by Alvan Clark, 5½ inches clear aperture ... The evening was transparent, and the Comet a truly magnificent object'. On 4 October it was 'a most magnificent object to the naked eye, not much inferior in brightness to Arcturus, which is about 5° from it'. He also observed it while he was travelling, and on 8 October it was 'a superb object during a long outside coach journey from Hereford to Clifford'.

On 18 June 1860 another Great Comet was discovered (C/1860 M1). It was near its best on 24 June, but Webb only made a couple of observations, on 27 June and 2 July , by which time it was fading.

Comet Thatcher (1861 G1) was discovered on 5 April and passed closest to Earth on 5 May. Webb observed it the next day – commenting that it was visible to the naked eye – and also on 8 and 9 May. This

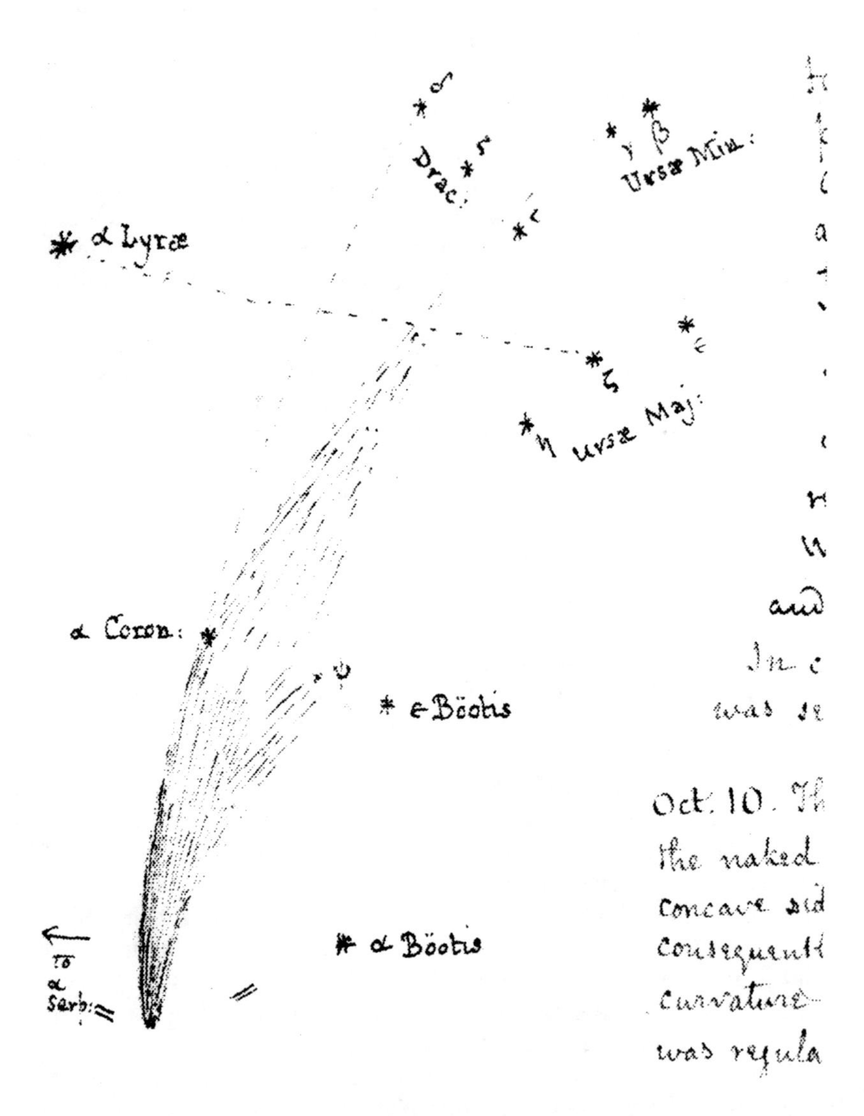

Donati's comet, 1858, in Webb's notebook, Volume 5, p. 204.

is an interesting comet, as it is the parent comet of the Lyrid meteor shower, and is unusual in that the period is very long at just over 400 years. (The meteor shower exhibits occasional outbursts, and I was fortunate enough to witness one of these from the South Atlantic in April 1982, when heading north from the Antarctic on board the RRS *Bransfield*. I was on board the research ship after a visit to the Antarctic, but we were in the South Atlantic due to events happening in the Falkland Islands – which, of course, have a strong link to another Thatcher.)

The Great Comet of 1861 (1861 J1) passed very close to the Earth (0.13 AU) and had a spectacular tail, through which the Earth may have passed. In many ways this must have been similar to comet Hyakutake of 1996, which passed 0.10 AU from the earth that March. It was a spectacular sight under dark skies, with the tail stretching halfway across the sky. The comet of 1861 clearly impressed Webb, and he devotes more space in his journal to this comet than to any other. He clearly made an independent discovery of the comet on 30 June, although Tebbutt had discovered it from Australia on 13 May. Webb recorded the discovery:

> The night being very clear, I went, about 22:30 GMT, not very inexact, to look out of the drawing-room window before going to bed, & saw a strange light from 5° to 10° above the NNW horizon, in a small vacancy between two trees – but for which clear space, & Mrs Webb's order that the shutters should be left open on account of the fineness of the evening, I should have wholly missed this night's observation.

His drawings made with the 5½-inch Clark show jets and dust shells very similar to those seen in comet Hale–Bopp. He also notes that several 'falling stars' (meteors) were noted during the observation. On 4 July, whilst describing the nucleus, he says: ' ... & preceded by a fan-shaped radiating sector, like an electric brush, strongly reminding me of some of my figures of Halley's Comet'. This suggests that Webb had observed Halley's Comet during its return of 1835, when it passed close to the Earth in October, and had seen the extraordinary jets recorded by Bessel, Smythe and Struve, amongst others, although it seems that his observations have not survived. There was much speculation as to whether the Earth had indeed passed through the comet's tail,[6] and after the event Webb recollected in his notes for 16 July:

> The clouds on the morning of July 1 ... consisted of curiously massed & contorted cirri ... the sky had certainly rather an unusual, but by no means a wonderful, much less an unprecedented appearance. The aspect

Comet Hyakutake, 1996.

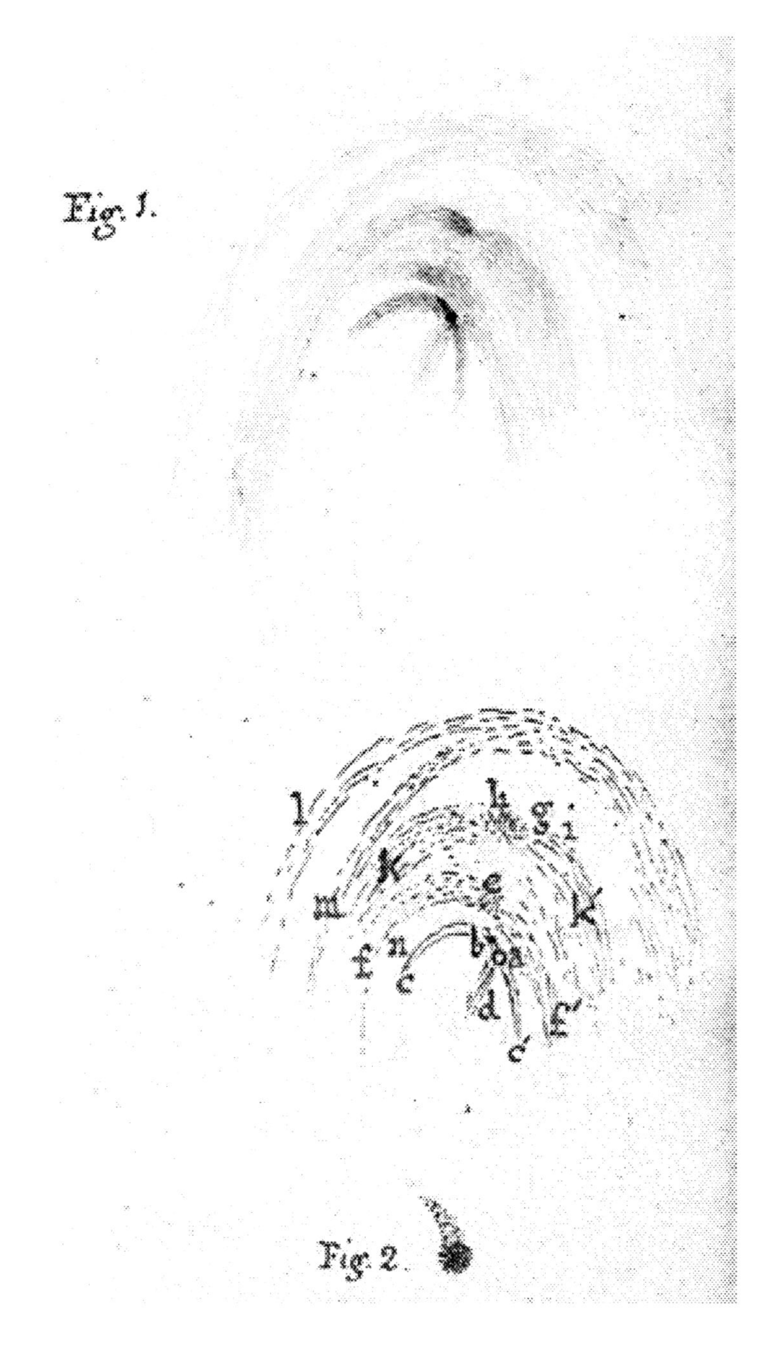

Drawings of the head of Donati's comet, 1858, in Webb's notebook, Volume 5, p. 207.

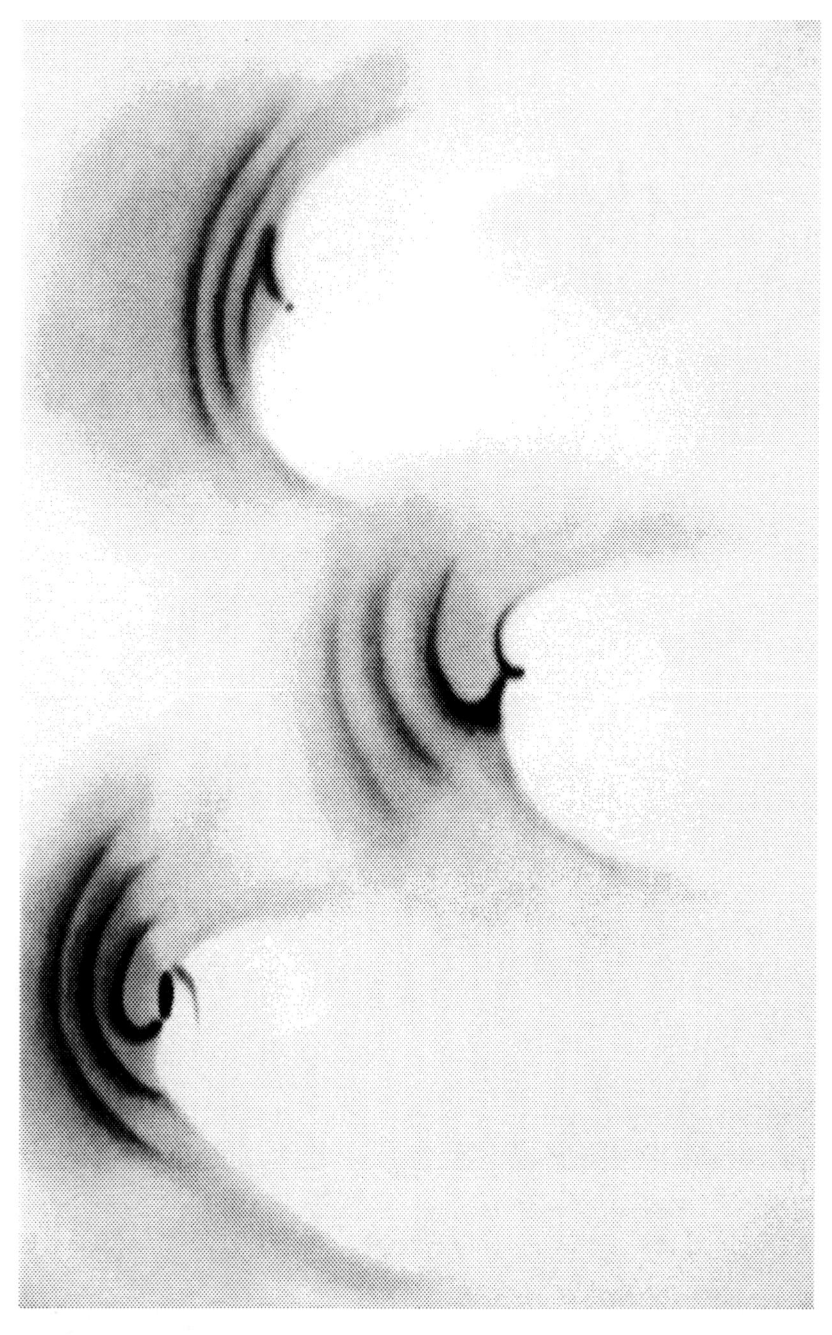

Drawings of the head of comet Hale–Bopp, 1997.

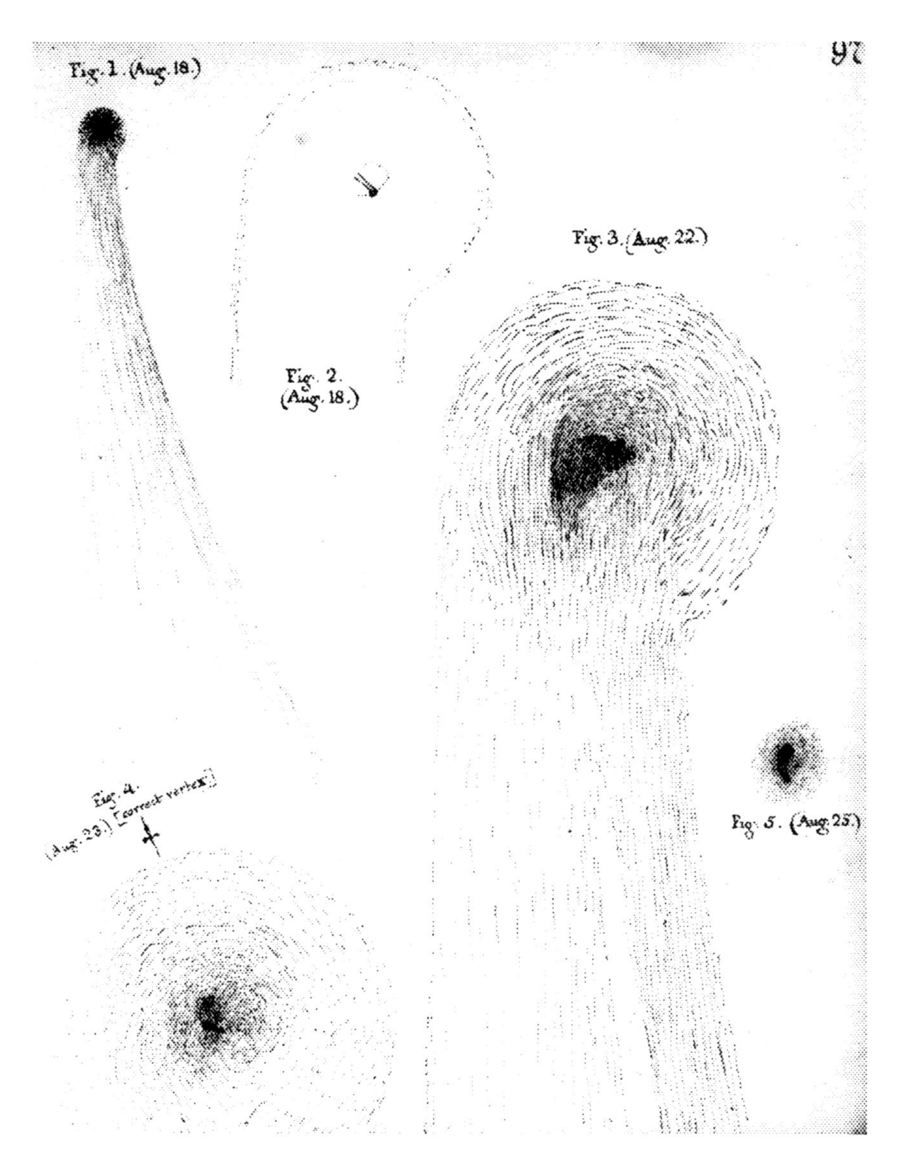

Drawings of the comet of 1862, in Webb's notebook, Volume 5, p. 97.

> of the cirri was very electrical, but no thundery weather followed; a showery season came on, & subsequently much very heavy rain.

These comments could be given greater credence if they had been made on 1 July, before any speculation of the tail passage had occurred.

Webb first observed the comet of July and August 1862 on 14 August after 'an extraordinarily cloudy season', although Mrs Webb had glimpsed it on 9 August. After this, however, he had a good run of observations, and followed the comet in detail until 28 August. He made detailed notes of a 'feather' or jet emanating from the nucleus. This comet was later identified as the parent comet of the Perseid meteors, 109P/Swift–Tuttle, and was predicted to return in about 120 years. The year 1982 came and went; and then when a new comet was found in 1992 it was linked with the comet of 1862, and also with comets seen in 1737 and earlier years.

Webb observed comet VI 1863 (most probably 1863 Y1), noting that it was barely visible to the naked eye and had a short tail. After this, Webb's observations of comets decreased, and only a few more are recorded in his journal, despite there being the opportunity.

On observing Encke's comet on 11 November 1871, Webb encountered a problem: 'Comet a mere cloud; no central condensation. It was very blotty [*sic*] in this region, over a chimney.' The comet has a period of 3.3 years and follows a cycle of apparitions, with those in the last quarter of the year being the most favourable for observation at British latitudes. Its appearance has not changed much over the years, and observations by the British Astronomical Association suggest no change in its absolute magnitude over the past sixty years.

The Great Comet of 1881 (1881 K1) was 'hidden by trees' until 9 July, when Webb first recorded an observation. He had clearly seen it earlier, as he commented that 'it had become comparatively feeble'. Other observations suggest that it had been 1st magnitude with a 20-degree tail towards the end of June and in early July.

Webb's final observation is that of the Great Comet of 1882 (1882 R1), which he and his wife observed from 3 October to 14 November. This was one of the most spectacular comets of the nineteenth century, and was a member of the Kreutz group of sungrazing comets. These pass a very small distance from the Sun, and as a consequence can become extremely bright with splendid tails. Modern observations with the SOHO spacecraft show that there is a continuous rain of small fragments falling in on the Sun, and we can only hope that a large object will appear and present us with a spectacle similar to that observed by the Reverend and Mrs Webb.

The Great Comet of 1882, in Webb's notebook, Volume 5.

Notes and references

1 T.W. Webb, 'Observations upon Natural Phenomena. Part III', MS in Hereford City Library.
2 RAS MS Webb 5.
3 G.W. Kronk, *Cometography, Vol. 2, 1800–1899* (Cambridge University Press, 2003).
4 T.W. Webb, *Monthly Notices of the Royal Astronomical Society*, **14** (June 1854), 218–24.
5 T.W. Webb, *Monthly Notices of the Royal Astronomical Society*, **19** (November 1858), 21–4.
6 T.W. Webb, 'The Earth in the Comet's Tail', *Intellectual Observer*, **1** (1862), 63–7.

Chapter 13

Webb's observations of the Sun

J.C.D. (Lou) Marsh

The solar phenomena are specially wonderful.

Although Webb was not primarily a solar observer, like many great amateurs of the nineteenth century his enthusiasm and dedication to all things astronomical led him to make a number of typically meticulous observations of the Sun, and ultimately, in 1885, to publish a small instructional textbook, *The Sun*,[1] of which a reviewer in *The Observatory* wrote:

> This neat little book is intended to be used in elementary instruction ... The book gives a clear and accurate account of the principal facts known about the Sun and the methods in which those facts have been ascertained ... We trust Mr. Webb's little book will meet with the success that the care bestowed upon it deserves.[2]

In *Celestial Objects for Common Telescopes* Webb devotes some thirty pages to observing the Sun, and in doing so covers, in depth, methods and techniques, sunspots, faculae, prominences, mottling of the surface, solar rotation and related observations of the annular eclipse of the Sun and the transit of Mercury.[3]

He also warns of the hazards of visual solar observing. He describes in some detail the methods extant at the time – glass filters of various colours and density, and unsilvered mirrors – and reveals some of the intriguing ideas devised by various observers. He refers to the technique adopted by William Herschel of using a trough containing a filtered mixture of ink and water, but offers no explanation of how this was used. He mentions in passing that Cooper, at Markree Castle in Ireland, used a drum of alum water and dark spectacles, whilst Merz of Munich was sufficiently forward-looking to adapt John Herschel's method of utilising a plane surface of unsilvered glass placed diagonally, by using four such surfaces and dark spectacles. Webb, however, strongly recommended observing by projection methods. Here again he discusses not only projection onto white card, but also describes a strange idea devised by William Noble, consisting of wet plaster of Paris spread smoothly on a piece of plate glass stuck to a firm support inside the base of a pasteboard cone blackened on the inside. Inevitably, the methods Webb advocated have now been superseded by more modern techniques such as the use of metallised plastic screens, but nevertheless many of his comments on filters of various colour and density are still valid today.

Webb's discussions on sunspots – 'dark spots', as he calls them – are precise and detailed. He describes the umbra and penumbra, notes the irregularity of the boundary, comments on the shape and intensity, and even mentions a bright bridge across a spot. His comments on sunspot groups include the elegant description 'a gregarious tendency is obvious', and he goes on to discuss the nature of such groups, particularly noting how spots often appear in pairs. He also noted their absence from the polar regions, and the way in which they varied with latitude.

His account of the great spot of June 1847 typifies his general approach:

> 1847. June 28, 6h.40m 5fA [5-foot achromatic], 144. Air rather unsteady. The spot, which must have been magnificent in the centre of the disc, is approaching the W limb. It was very dark and more irregular in its outline than I could succeed in delineating. The umbra was so evidently darker towards its outer edge that it gave the effect of a more luminous ring surrounding the nucleus. Towards the SE it had been compressed and made to deviate from its irregular form by another small umbra and here the bright disc appeared in contact with the more luminous portion adjacent to the nucleus. A very extensive branching facula surrounded the whole phenomenon. Two fainter spots, considerably smaller appeared at a little distance to the S and two more were traversing the disc more towards its centre.[4]

The following day he recorded:

> 1847 June 29. 5h 10 min 5fA, 112. air very unfavourable. The spot has approached so near the limb that it is but imperfectly seen; but the side of the umbra nearest the limb appears, I think, broader than the others as, if it

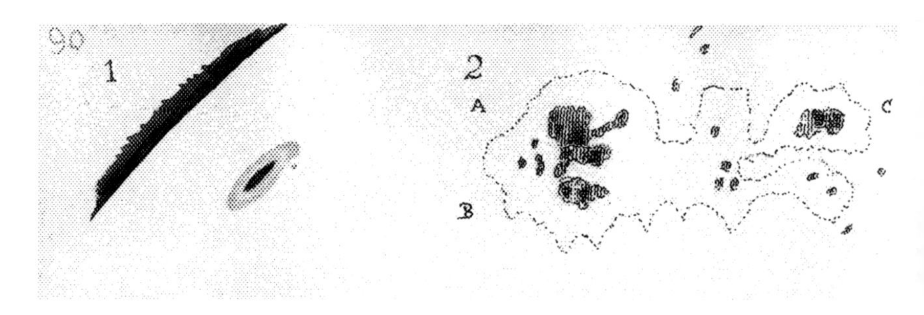

Diagram of a sunspot, in Webb's notebook, Volume 5.

> is depressed beneath the surface, it should do. The difference is however not considerable and the whole spot has the aspect of being very shallow.

The nature of sunspots was then virtually unknown, and the faint Wilson effect – which has to do with the appearance of sunspots as they approach the limb, giving the impression of a shallow depression – was clearly recorded by Webb.

A further detailed description of a group of spots observed on 16 January 1855 is also worth noting:

> The great spot, now not very far from the sun's centre, but above it, was observed with 144 and a bluish grey glass. The air was not steady but definition was not too bad. There were three great nuclei A, B, C, included in a common umbra beyond which, however, several small spots were to be found. The umbra was very clearly defined, bright on the N side encroached in pointed promontories upon the surrounding surface: it was mottled with light and shade in several places including specks of considerable brightness especially a lengthened spot close to the S edge of spot B between it and A which seemed as brilliant on the mottled surface as the unspotted disc. A was divided in two by a very narrow, slightly curved bridge of light, which, during the progress of the observation from about 1h 30m to 2h., appeared to be evidently melting away. The dark ground of the three nuclei was evidently overspread in parts with a faint light, through which when the air was more favourable, very deep openings may be seen. In A there were three of these black spots – in B two and in C two or three – but each nucleus differed from the others in the intensity of this faint light.

He went on to describe these in some detail, and concluded: 'The whole impression was that of increasing depth, passing through stratum after stratum of luminous matter which seemed to dissolve into deepest darkness at last.'

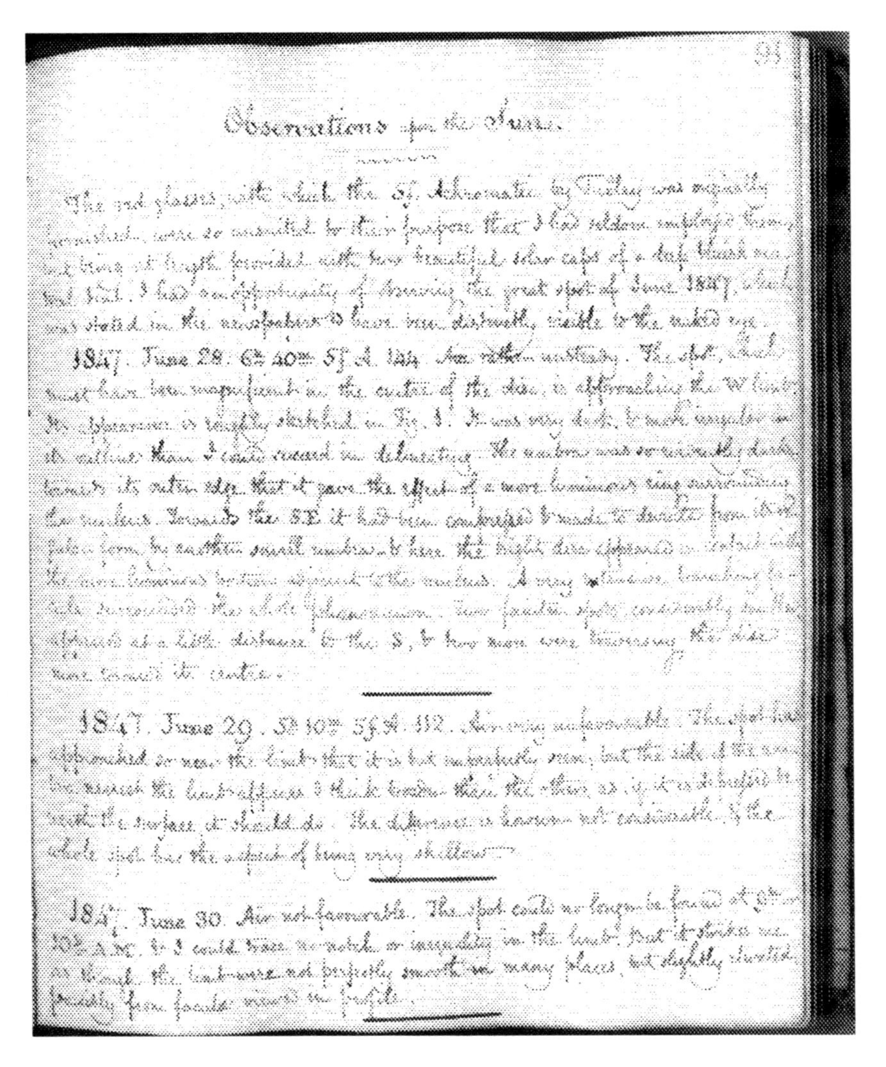

91

Observations upon the Sun.

The red glasses, with which the 5f. Achromatic by Tulley was originally furnished, were so unsuited to their purpose that I had seldom employed them; but being at length provided with two beautiful solar caps of a deep [illegible] I had an opportunity of observing the great spot of June 1847, which was stated in the newspapers to have been distinctly visible to the naked eye.

1847. June 28. 6h 40m 5f. A. 144. Air rather unsteady. The spot, which must have been magnificent in the centre of the disc, is approaching the W limb. Its appearance is roughly sketched in Fig. 1. It was very dark, & much irregular in its outline than I could record in delineating. The umbra was so remarkably dark towards its outer edge that it gave the effect of a more luminous ring surrounding the nucleus. Towards the SE it had been contracted & made to deviate from its [illegible] form by another small umbra & here the bright disc appeared [illegible] the more luminous portion adjacent to the nucleus. A very [illegible] surrounded the whole phenomenon. Two fainter spots considerably [illegible] appeared at a little distance to the S, & two more were traversing the disc more towards its centre.

1847. June 29. 5h 10m 5f A. 112. Air very unfavourable. The spot had approached so near the limb that it is but imperfectly seen; but the side of the umbra nearest the limb appears I think broader than the other, as if [illegible] with the surface it should do. The difference is however not considerable, & the whole spot has the aspect of being very shallow.

1847. June 30. Air not favourable. The spot could no longer be found at 9h or 10h A.M. & I could trace no notch or irregularity in the limb. But it struck me as though the limb were not perfectly smooth in many places, but slightly dented, possibly from faculae viewed in profile.

Observations upon the Sun, Webb's notebook, Volume 5, p. 91.

On 1 September 1855 Webb observed that the disc was perfectly free from spots but the mottled or granular appearance was very evident; and on 5 September the mottling was evident over the whole disc. He became very interested in the mottling, as, he said, Sir John Herschel thought it could not be seen with achromats. However:

> On 1855 Oct 16 about 11.30 I viewed the sun attentively. With a deep red glass it might be seen, but not with comfort, on account of the intense

> heat. With a pale bluish grey it could be well seen with 80 and 144 but 250 did not show it so readily.

Further observations of the mottling were recorded in November 1855 and May 1856. On 10 May 1856 there is a clear reference to limb darkening:

> Air fine but sun low ... No spot: a general diffused equable and minute mottling and much diminution of light towards the limb, which through a neutral blue glass appeared shaded with a faint brown hue.

Webb also commented on the faculae (bright streaks), which, he said, are to be looked for in the spot-bearing regions, but only near the limb, and are irregular, curved and branching. He agreed with W.R. Dawes' conclusion that they are ridges in the photosphere which, near the limb of the Sun, appear as projections above the limb. (It should be remembered that there was then no way of observing prominences except during a total eclipse of the Sun.)

On 15 March 1858 Webb observed an annular eclipse of the Sun; but it was

> ... only a source of disappointment. The remaining portion of solar light was exceedingly small yet the darkness which was only conspicuous for a few minutes, did not appear to exceed that of about an hour after sunset; and the sky was so entirely covered at The Moor where we were that not a trace of the sun could be seen. At the greatest obscuration the clouds became somewhat reddish and the earth had a strange greenish and reddish hue, very unnatural but not more so than that of a thunderstorm. A little before this took place the cocks crewed a great deal and rooks and small birds were seen going to roost.

Again disappointment overshadows his observation of the transit of Mercury on 12 November 1861:

> The sun rose in a sky so full of streaky haze that though the planet was visible with T65 of Alvin [*sic*] Clark's 7F achromatic, I did not recognise it, having mistaken the position upon the disc and imagined it was merely an ordinary spot. Clouds very soon came on: when they had departed they left a sky of the same unfavourable character, in which the last half hour and the egress were visible with 170. Definition was so bad that no details could be satisfactorily made out.

In the last of Webb's available manuscript notes on the Sun (now in the archives of the Royal Astronomical Society) he describes the partial solar eclipse of 22 December 1870:

> The great eclipse of 0.8 very much clouded but occasionally pretty well seen and considering the very low altitude sharp with 65 on a 9 inch mirror reduced to about 4 inches minus flat [secondary mirror]. Limb of moon rather not much sharper than that of the sun. Moon though darker than sky around the sun.

Webb pointed out that partial solar eclipses are seldom of much interest except for the occasional projection of mountains on the Moon's limb, and although he commented on the beauty of a total eclipse, there is nothing to indicate that he ever saw one. He said that they are very rare – and also noted that the next opportunity to see a transit of Venus would be in 2004.

Finally, Webb devotes six pages of *Celestial Objects* to a subject which nowadays has curiosity value only, but during the nineteenth century was linked to a major scientific question: the transit of unknown dark bodies across the disc of the Sun. This subject had relevance to the anomalous advance in the perihelion of Mercury – a problem that induced a belief in the possible existence of an undiscovered body in orbit between Mercury and the Sun – the so-called intra-mercurial planet hypothesis.[5] Other than in a passing reference to the Watson–Swift observations of 1878, Webb does not especially specify this problem (which was ultimately solved by Albert Einstein). However, his knowledge of the day-star, and what might lie in its immediate environs, is articulated with a confidence that is based on a thorough knowledge of the literature of his time, confirming his erudition and understanding of the practicalities of solar observing.

Notes and references

1 T.W. Webb, *The Sun: A Familiar Description of his Phænomena* (Longmans, London, 1885).
2 Review of *The Sun*, in *The Observatory*, no. 97 (1885), 181.
3 T.W. Webb, *Celestial Objects for Common Telescopes*, fourth edition (Longmans, Green and Co., London, 1881), pp. 19–45.
4 RAS MS Webb 5, pp. 90–6, from which subsequent quotations are taken.
5 R. Baum and W. Sheehan, *In Search of Planet Vulcan* (Plenum, New York, 1997).

Chapter 14

Double stars, garnets and rubies

Robert W. Argyle

Here indeed is a field where enterprize cannot be thrown away, nor perseverance fail of its reward

Double stars

Webb's observations of double stars began in earnest when he undertook to reobserve all the objects in Smyth's Bedford Catalogue (the second volume of *A Cycle of Celestial Objects*[1]), using his 3.7-inch Tulley refractor. In his observing notes dedicated to this work, Webb wrote out a summary of each entry in the Bedford Catalogue on the *verso* side of the page, and then added his own observations on the *recto* side, although he did not observe every object, due to the small aperture used (Smyth used a 5.9-inch Tulley refractor to compile the catalogue). The project began in 1847 and took nine years to complete.

The Bedford Catalogue contains 850 objects, of which ninety-eight are nebulae, seventy-two are clusters and the remainder double stars, although Smyth divided them into stars and comites (S c), double stars (D), binary stars (B), triple stars (T), quadruple stars (Q) and multiple stars (M).

The apparent relative position of double stars in a telescope eyepiece is defined by two quantities: the position angle (p.a.), which is the orientation of the line between the two stars measured from north (0°) and increasing through east (90°), south (180°) and west (270°); and angular separation, usually expressed in arcseconds ("). In the case of

Webb's 3.7-inch refractor he could expect to just divide two equally bright stars separated by just over 1 arcsecond. S.W. Burnham, with the 36-inch refractor at Lick Observatory in California, could resolve pairs separated by about 0.1 arcsecond. (By comparison, the keen unaided eye can resolve bright double stars such as the components of ε Lyrae, which are separated by 3.5 arcminutes). Rather than give position angle and separation as numerical estimates, Webb preferred to record that the fainter star was south preceding (*sp*) or north following (*nf*) the brighter star, and was content to say, for example, that a pair was 'well-separated'. Observers such as Dawes used micrometers – essentially a grid of fixed and moveable wires, the orientation of which was measured precisely with a brass circle to measure much more accurately the relative positions of double stars. In the case of binary stars, when the two stars are gravitationally connected rather than just coincidentally in the line of sight, the fainter star may be seen to revolve around the brighter in a few tens of years, in pairs such as ζ Cancri and ξ Ursae Majoris. (In reality, of course, both stars revolve around the common centre of gravity (the barycentre) – the brighter star being used as a convenient reference point. This means, however, that only the total mass of the two stars can be derived from the relative positional measures, and the accuracy of this figure depends very much on the accuracy of the relative positions being supplied.)

Webb's comments on the observations of close double stars during this period lead one to conclude that the performance of the 3.7-inch was fully commensurate with expectations. The best results were ζ Cancri (1".0, 1849), 'perhaps elongated', and Σ1517 Leonis (1".0, 1851), 'a little elongated', whilst Σ1126 Canis Minoris (1".3, 1856) had 'discs in contact'.

The 9⅓-inch With–Berthon reflector was also obviously of high quality. One of the regular test objects used by With and Calver was γ^2 Andromedae. The 8½-inch mirrors of both makers were guaranteed to divide the pair, at a time when the separation was about 0".6. Webb also noted, in 1878, that he was able to suspect a division of ω Leonis, then at 0".52, and to divide η Coronae Borealis at 0".55.

His entries in the observing book are each prefaced with a short note about the sky conditions, and each individual observation is numbered, making future references to the same object much easier to trace. They are in a neat shorthand style, thus:

> 1851 Feb 22 A very good night
> 636 84 P.V. (Or:)(F ψ^1)(144) Very minute but certain. I think white and bluish.

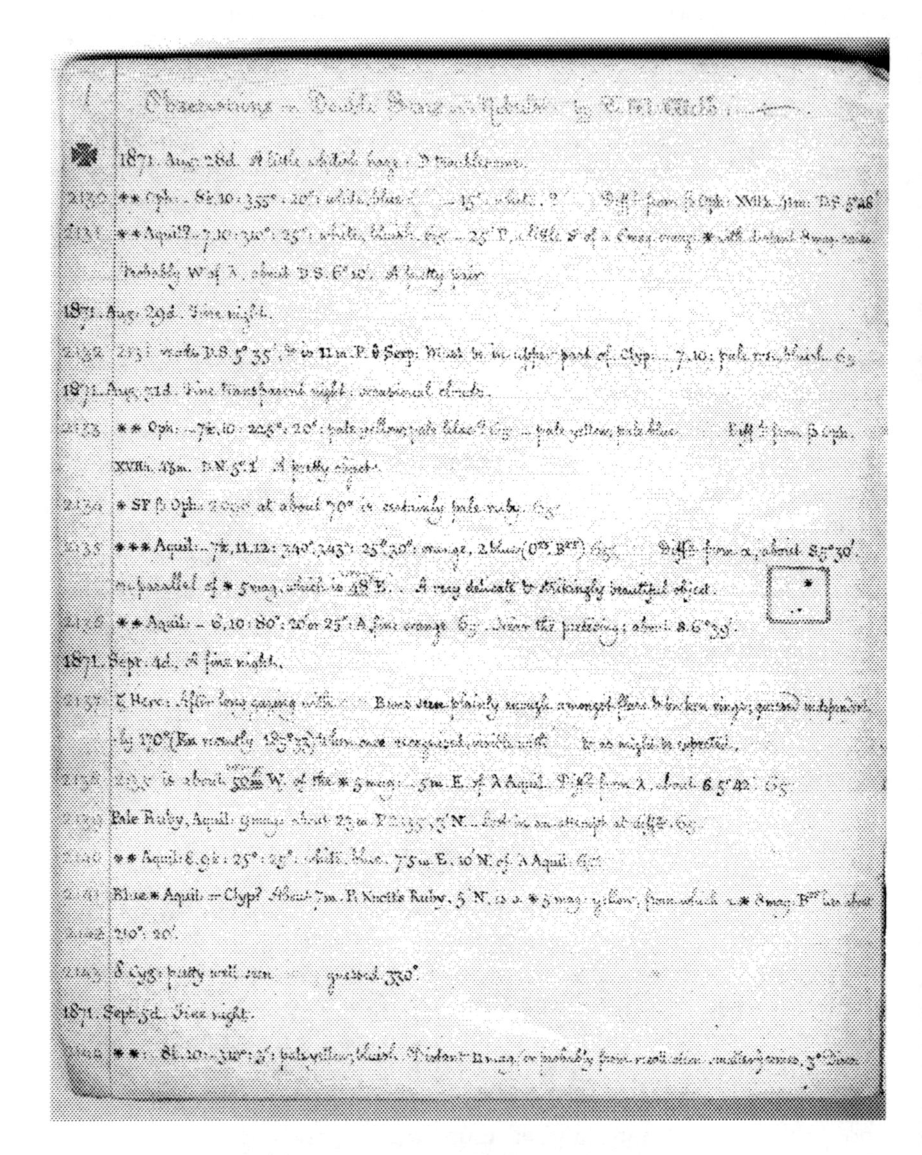

The first page of Webb's last observing notebook, Volume 5, which was started on 28 August 1871. The observation numbers were left out and filled in later in red, as were the magnifications employed, and although there appears to be a faint pencilled entry underneath this in many cases, it is not always overwritten, as in observation 2137 for example. The entries are short and succinct, and the state of the sky is usually dealt with in a few words.

(1885 Mar. 7d.)

rough est. 310°. 25" or 30". Fine object for small telescope.

Cluster. Mon. While taking long sweeps to differentiate the preceding, I came upon a noble bright cluster far W, & ±1° 10' S of it; many 9m. stars grouped towards the middle in a kind of festoon, which misled me into the absurd mistake that I was looking at M35! So much was I impressed with this that I looked for, but in vain, a red star which I believed there was in M35: but I noted a solitary star near the centre of the curve. I had to leave the telescope. & immediately after discovering my mistake, & on going back, having turned it in its cradle, could find neither this, nor the preceding object again, but noticed that the vernier, I believe on the ✱✱, stood below 5° – I think near 5° 30'. I had not then time for further sweeping.

1885 Mar 12d. Haze & frequent clouds – 9⅔m.

3459 A clouded & very bad differentiation of 3458 led me to believe that it is G.C. 1657 (Mon.)

1885 Mar. 13d. A slight haze & most [illegible] [illegible] [illegible] [illegible] [illegible] wind making it difficult to read divisions & once blowing out the candle. – 9⅔m.

3461 Pale Ruby. Mon. [illegible] [illegible] F [illegible] joining Sirius & Procyon, & [illegible] from Sirius [illegible] 5° 23'. [illegible] [illegible] [illegible] [illegible] [illegible] (very rough [illegible]) 240°. [illegible] [illegible] [illegible] [illegible]

3462 [illegible] W, which I could not identify, 2 or 3' N of the star.

3463 S Or: very small. I thought about 2 m. below e: e 0.5 m. [illegible]

Omitted in its place above.

1885 Mar. 14d. Sharp discs through a restless flutter. – 9⅔m.

3460 [illegible] Dr Copeland having mentioned perfect definition, with 960 on the 15 in. O.G. with which the star was merely "fluffy," but the close pair never [illegible] the least confused, I tried it with [illegible], & [illegible], & was surprised to find it so very easy – two round hard discs with a perfectly black division, ½ or ⅔ of either: rough est. ±45°

The last observing entries made by Webb, 7 March 1885. The very last observation – number 3463 – is of his own discovery, the variable star S Orionis. Even though he was 78 years old, his writing shows no lack of vigour.

This is observation no. 636 of the double star P (Piazzi) V 84 Orionis (Σ712). The star follows ψ^1 Orionis, and was observed with the 3.7-inch Tulley at x144 magnification.

Pairs were usually located by referring to the nearest bright star or stars in terms of offsets in Right Ascension and declination, or by a small drawing showing the local neighbourhood of field stars; but sometimes he used features in his star atlas. He began with Alexander Jamieson's atlas, then acquired the maps prepared by W.R. Dawes for the Society for the Diffusion of Useful Knowledge, and later used Richard Proctor's atlas. Examples are an observation in the northern half of Perseus, 'In Cap: medusae' (in the Medusa's head, which appears on Jamieson's map); and 'seems to be near the star N. of the first letter S in the word SAGITTARIUS (Jamieson)'.

Webb adopted the constellations as given in Jamieson, which differ from present-day versions. For example, Antinous is now part of Vulpecula, Clypeus Sobieskii (Sobieski's Shield) is now Scutum, and Taurus Poniatowski (Poniatowski's Bull) no longer exists. Jameson lists 112 constellations, of which some sixty-seven are recognisable on the star maps of today. The Milky Way (Via Lactea) is treated as a constellation.

It is apparent from Webb's notebooks that he often came across pairs that he could not at first identify, given his sources of reference, and he was not privy (at least initially) to F.G.W. Struve's great Dorpat catalogue, *Mensurae Micrometricae*,[2] containing more than 3,000 double stars, although he was later able to go back to some of his earlier observations and append the Struve (Σ) numbers. A number of pairs were new. The *Washington Double Star Catalog* (WDS) assigns ten pairs to Webb, and these are tabulated below, together with a discussion of each pair.

Nearly all of Webb's observations are given as eye estimates – perhaps simply '*nf*' (north following) – though he would sometimes hazard an estimate of position angle and separation. He did not own a micrometer, but in his later observing notes he describes the use of a 'rough micrometer' which appears to have consisted of a wire (or bar) in the eyepiece and an angular scale, on cardboard, around the eyepiece. The bar is first mentioned on 7 February 1878, and subsequent observations appear to put its angular diameter at about 30 arcseconds. In 1882 Webb recorded that he had a 'rough micrometer fitted up by Parkes' which used a Kellner eyepiece and appeared to have some form of illumination. This device (which he did not use much) would have allowed him to make estimates of separation, from the thickness of the wire, and position angle, from its orientation. He also used the bar to hide bright stars, allowing him to check that red stars were not that colour due to contrast.

Webb was certainly aware of the micrometric work being carried out by Dawes, Smyth and others, and he would certainly have been able to afford a large telescope and micrometer had he so wished. He may well have felt that his eyes were not keen enough to make micrometric measures, and the outstanding work by his friend Dawes may well have decided him against this line of research. Perhaps, too, he had insufficient time for such work, due to his clerical duties. What can be concluded in reading the notes, however, is that whilst Webb would follow up anything that caught his attention, he was happiest just wandering through the heavens. In *Celestial Objects*, in the section for Cygnus, he writes:

> I had at one time projected a survey of the wonders of this region with a sweeping power, but want of leisure, an unsuitable mounting, and the astonishing profusion of magnificence, combined to render this task hopeless for me, which, I trust, may be carried through by some future observer.

Webb's sources of reference

Alexander Jamieson, LLD, wrote books on the construction of maps (1814), fluid mechanics (1837), and a volume on the grammar of logic and intellectual philosophy. In the preface to his star atlas[3] he writes: 'If there be any merit in reviving what the immortal Flamsteed first projected, I trust I have earned it.' It seems clear from this that the star positions are taken from Flamsteed's *Historiae Coelestis Britannicae*,[4] published in 1725 but precessed to 1820. The book consists of twenty-nine star maps and a plate illustrating the Moon, Venus, Jupiter and Saturn. There are lists of the principal stars in each constellation, computed for 1820. The constellations number 112, including many small additional constellations which no longer exist, such as Telescopium Herschelii (Herschel's Telescope) – apparently added by Jamieson to honour the greatest living astronomer. Jamieson also transformed Le Monnier's constellation L'Ermite (Oiseau) into Noctua – a bird which appears in ancient Egyptian monuments. Perhaps in gratitude to – or to gain the approval of – King George IV, to whom the atlas is dedicated, the figure of Virgo is represented by Princess Charlotte.

In the star catalogue that accompanies the maps, positions are given to the nearest minute of Right Ascension in hours and minutes, but to the nearest arcsecond in angular form. The Declination is also given to the nearest arcsecond. Magnitudes are also given, the faintest being 7.8, and a few stars at 7.0 and 6.7. Magnitude 1 is applied broadly to stars ranging from Sirius to β Leo, which differ by almost 3 magnitudes.

Jamieson may have been a rather abrasive character. In the preface he talks of 'the rebuffs which came from quarters where I least expected them', and having been elected a Fellow of the Royal Astronomical Society on 12 May 1826 he resigned on 6 February 1833.

In many observations, Webb refers to the star maps produced by the Society for the Diffusion of Useful Knowledge. This organisation was formed in 1826 at a 'meeting of gentlemen' convened by Mr (later Lord) Henry Brougham, with the aim of 'promoting the composition, publication and distribution of elementary works upon all branches of useful knowledge'. By November 1841 this had led to the publication of 168 volumes covering many areas of knowledge, including twenty-one volumes of the *Penny Cyclopaedia* and a set of star maps. The Society closed in 1848.

The SDUK produced several versions of the star maps in gnomonic projection. The first version, published in 1833, consists of six folded

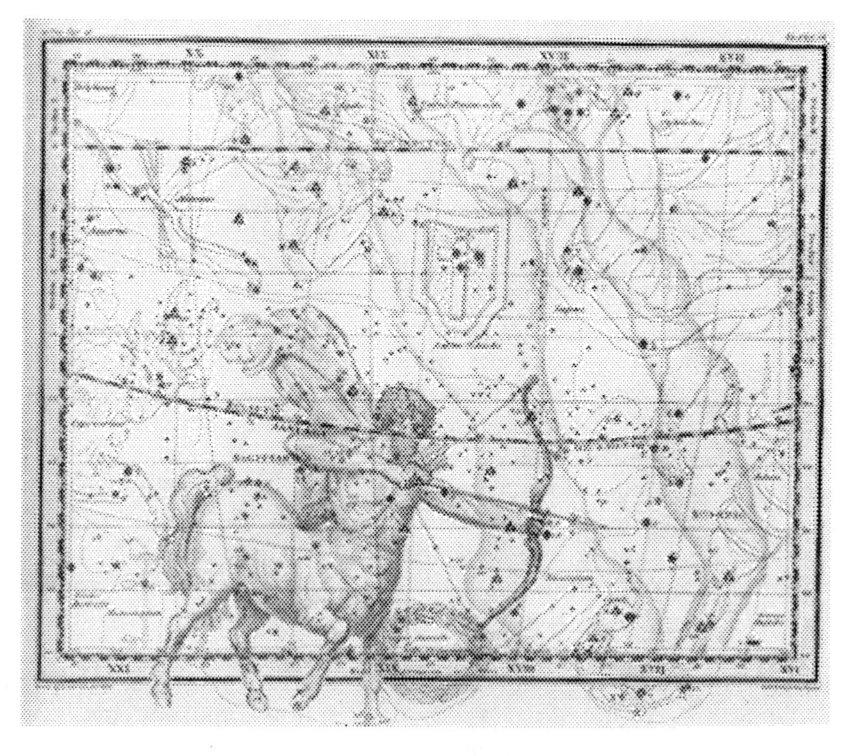

A typical page in Alexander Jamieson's star atlas.

maps, 25.2 x 24.4 inches. The key shows stars marked from magnitudes 1 to 9. A revised version containing six maps, with stars laid down according to gnomonic projection, was produced by W.R. Dawes in 1844. This was published in two editions: a hand-coloured version on cloth, 10.2 x 11 inches, costing 6 shillings; and a slightly larger uncoloured copy on paper, costing 3 shillings. It is clear from Webb's comments in his observing books that he had the large version.

Webb later became more critical of the distortion at the edges of the maps he was using, and then suggested that the smaller of two atlases by R.A. Proctor was more useful.[5] This contains six circular maps on equidistant projection for each hemisphere, and does not have the constellation figures superimposed. As his source, Proctor used the catalogue of the British Association for the Advancement of Science (compiled for the BAAS by Francis Baily, who carried out most of the work but died before it was published).[6] From the 8,377 stars in the catalogue, Proctor plots 2,984 stars in the northern hemisphere and 4,159 in the southern hemisphere down to magnitude 7 (which prompted him to wonder why there was such a large difference). Double stars, variable stars and nebulae are also plotted.

Relations with contemporary double-star observers

Webb maintained contact with the most active double-star observers of the day – Dawes, Dembowski and Knott amongst others – and in 1860 he paid a visit to Hartwell, the home of Dr Lee, to where Smyth's Tulley refractor had been transferred in 1839 when the Bedford observatory was dismantled. In 1859 Webb had acquired a 5½-inch Clark refractor, and having using Smyth's telescope and examined ζ Cancri on the evening of 1 May he reported that 'it appeared inferior to my own'.

It was, however, the correspondence between Webb and Burnham, and the influence that *Celestial Objects* had on the work of the great American double-star observer, that was most notable. Sherburne Wesley Burnham (1838–1921) had begun serious double-star work in 1870, and had bought a superb 6-inch Clark refractor with which he was making numerous discoveries. Early on, Webb had expressed his doubts whether the rate of Burnham's double-star discoveries could be maintained, but Burnham demonstrated that there were many more pairs to be discovered; it did not even need a large telescope, but very keen eyesight and excellent optics were an advantage. On the appearance of the second edition of *Celestial Objects* in 1873, Burnham wrote to the *English Mechanic*, welcoming the new edition, and admitting: 'I rarely go to use the telescope without taking three books: a note-book, Webb's *Celestial Objects*, and Proctor's *Star Atlas*.' He also said that he would prefer to have Webb's book than a copy of Smyth's *Cycle*.[7]

Burnham went on to use the biggest telescopes in the world during the next forty years, and his first major achievement was his *General Catalogue of 1,290 Double Stars*, published in 1900.[8] Even here he did not forget the influence that Webb's book had had on him. In the Introduction he says:

> At the beginning of the use of the 6-inch telescope my library, so far as the subject of double stars was concerned, was principally confined to the first edition of Webb's *Celestial Objects for Common Telescopes*, and I wish here to record my great indebtedness to this most admirable and really indispensable book. It was of great assistance to me at the time, and it has never ceased to be a valuable and convenient work for useful reference.

Double stars discovered by Webb

The WDS lists the following pairs and multiples under Webb's name. In each case the table, with Right Ascension and declination for J2000, contains the latest available measurement of position angle and separation. BD is W. Argelander's *Bonner Durchmusterung*.

RA	Dec	Pair	Date	PA°	Sep"	Visual mags.		Spectrum	BD
03219	+4904	WEB 1	2002	328	213.3	5.98	9.64	F6V	+48 893
03427	+5958	WEB 2AB	1913	95	21.4	5.72	13.8	K4II	+59 699
		WEB 2AC	1913	300	34.9	5.72	13.0		
		WEB 2AD	2002	36	55.3	5.93	8.45		+59 700
		WEB 2AE	1909	161	168.2	6.0	10.8		
05231	+0117	WEB 3	2003	10	62.2	7.29	10.45	K2	+01 992
05290	–0442	WEB 4	2001	233	47.6	9.93	10.09	M7e	–04 1145
06097	+4308	WEB 5	2002	216	43.5	7.10	9.19	A0	+43 1466
16354	+1703	WEB 6	2001	359	157.0	6.41	7.26	A2V	+17 3053
18448	+4458	WEB 7	2003	49	11.4	9.55	10.35	A5	+44 2985
19520	+1018	WEB 8	1920	48	11.8	10.5	11.5		
20007	+3635	WEB 9AB	2004	203	71.1	6.69	8.97	B6IV	+36 3816
		WEB 9AC	2004	220	82.1	6.68	9.6		
		WEB 9BC	1999	277	26.3	8.97	9.84	G	+36 3815
		WEB 9CD	1998	326	14.1	10.1	10.6	A2	+36 3814
23386	+4441	WEB10AB	2004	304	125.2	8.30	8.76	F5	+43 4515
		WEB10AC	1991	249	121.5	8.30	9.99		+43 4513
		WEB10BC	2004	185	114.4	9.0	9.9	A0	+43 4514
		WEB10BD	2004	253	105.4	8.76	10.31		

WEB 1 Included in the first and second editions of *Celestial Objects*, this pair appears as P III 37 with the following description: '6m. orange – near alpha (Persei) s. a little p. – has a fine blue companion in a beautiful field.' In the fifth edition the description is similar, but the pair is correctly identified as P III 28. It was first measured by W.S. Franks in 1914. From 1913 to 2002 the distance had increased by 8". It was first observed by Webb on 13 November 1850.

WEB 2 First noted on 2 March 1854, this is P III 97 with colours of orange and blue, as noted by G. Knott (RAS *Memoirs*, **43** (1875), 75). Distance +2".7 in 21 years. AC little change in 10 years; AD –1".1 in 192 years; AE +1".8 in 120 years. Stars B, C and E found by T.H.E.C. Espin. G. Piazzi (*Praecipuarum Stellarum Inerrantium Positiones Mediae* (Palermo, 1814)) noted the fainter star 'Ad Boream sequitur alia Summe exigua'. (It is followed to the north by another much smaller star, but he did not measure it.)

WEB 3 This is given as P V 67 and is first noted by Webb on 20 January 1865. In his paper, Franks (RAS *Monthly Notices*, **74** (1914), 517) gives colours of pale orange and blue. Little change. Piazzi (ibid.) says 'Ferem in eodem verticali binae Sequuntur vicinior 9 mag, altera 8ae, utraque ad Austrum'. (They are followed in the same vertical – closer 9th mag, another 8th and another to the south.)

WEB 4 S Orionis. First measured by Burnham with the 18.5-inch Clark refractor at Dearborn Observatory. Distance +0".9 and angle increasing by 6° in 122 years. Faint comites first recorded by Webb on 27 December 1869.

WEB 5 Espin (RAS *Monthly Notices*, **74** (1914), 244) quotes Webb as the source in his list of published observations. The original observation was made on 6 April 1878, when Webb estimated 220°, 35" and magnitudes 7.5 and 9.5. Distance –0".5 in 108 years.

WEB 6 On 20 August 1855 Webb noted two pairs in the region of κ Herculis and ω Serpentis which he could not identify. On returning to the area on 12 October he noted another pair for which he noted a position at 16h 22m +16° 30', but could not say if this corresponded to either of the earlier observations. This latter pair forms P VI 125/126, and appears in the first edition of *Celestial Objects* with an indication that it is not in the Bedford Catalogue. The separation has increased by 0".8 over 102 years.

WEB 7 Webb happened on this pair on 24 July 1875, whilst sweeping near δ Cygni (Espin, RAS *Monthly Notices*, **78** (1918), 189).

He observed it again on 24 August, noting that it 'is probably not in Σ at all'. Distance +0".4 in 110 years.

WEB 8 E. Doolittle (*Pub. Univ. Pennsylvania*, **4**, Part 1 (1915)) says: 'Noticed by Webb (*Cycle of Celestial Objects*) p.29', but this should be *Celestial Objects for Common Telescopes* (fifth edition). Webb mentions it for the first time on 28 August 1865 as a 'minute pair p. a little s. of P XIX 307' (Σ2590).

WEB 9 First noted by Webb on 12 September 1878, due to the existence of a red star in a pretty group of four stars. Webb estimated the distance and angle between the brightest star (mag. 7) and the red star (mag. 10) about 207° and 105", but the WDS gives 201° and 71". Further observations (3123, 3141 and 3144) refine the position to 19h 58m.5 +36° 10' (1880) – equivalent to 20h 03m +36° 30' (2000), some distance away from the WDS position (20h 00m.7 +36° 35'), but there is nothing else in the field, and comparison of the blue and red plates of the *Palomar Observatory Sky Survey* shows that star B is very red. There is little change in the relative positions.

WEB 10 First observed by Webb on 13 October 1871 (2197): '**** And: – 9.10 mags. a very curious group, nearly an equilateral rhombus, XXIIIh.30 or 40m :N 44°.' Franks first measured it in 1919. AB unchanged; AC +8".6 in 104 years; BC +1".3 in 110 years; BD +1".0 in 110 years; CD +0".9 in 104 years.

Red stars and variable stars

Webb was not a particularly avid observer of variable stars. His later notebooks show only occasional observations of the brightest long-period stars such as χ Cyg, Mira (o Cet), R Leo, and a few others. Later in his career, however, he took an abiding interest in red stars, and made many observations, including discovering a number of them during sweeps in Milky Way fields. This was inspired by the list of 280 stars published by Schjellerup in *Astronomische Nachtrichten* in 1866.[9]

Some years later Webb suggested to John Birmingham that the latter should undertake a revision of this list, which Birmingham duly did. Birmingham's catalogue of 658 red and orange stars[10] also owes a debt to Webb in that many of the observations were supplied by him. Whilst it is not clear who first noted these stars, Webb's name is the first name in the notes on thirty-two stars, and his observations of another fifty-five are acknowledged.

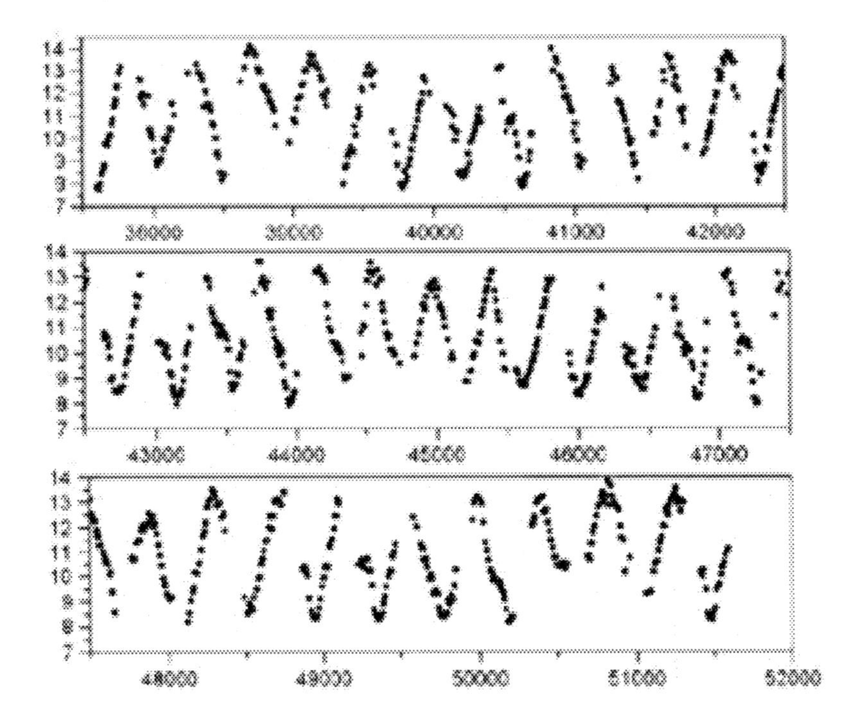

The light curve of the variable stars S Orionis between Julian Date 2437600 and 2451600, as presented by P. Merchan Benitez and M. Jurado Vargas, based on observations by members of the AAVSO, VSOLJ and AFOVE. Visual magnitude is plotted against Modified Julian Date. (Reproduced from *Astronomy and Astrophysics*, **386** (2002), p. 245.)

Webb made two significant discoveries. The first was the planetary nebula in Cygnus now known as NGC 7027, which was found on 14 November 1879;[11] but perhaps his most well-known discovery was made on Christmas Day 1869, when he came across a 'decided carmine star in Orion'. Reobserving it two nights later he noted that it formed part of a triple star in the eyepiece of his 9⅓-inch With–Berthon reflector. The other two components were located at p.a. 230° and 340°. By 26 January 1870 he was in no doubt that it had increased in brightness, having compared it with the two wide companion stars. This star – later designated S Orionis – and the nearer of the two companions in this coarse triple star form the pair WEB 4 in the *Washington Double Star Catalog*.

In late 1870 Webb was again observing his carmine star. On 24 December, he noted, it 'no doubt remains that it is a *new variable*'; and some 12 months later he first referred to it as S Orionis. In 1870 a

The wide multiple star WEB 10, 17 February 1998. Image obtained with a 100-second exposure using a 5-inch Astrophysics refractor at f/6 and an ST7 CCD camera.

summary of the discovery and subsequent observations was published in the RAS *Monthly Notices*.[12] This star meant a great deal to him, and he observed it without fail every subsequent winter for the rest of his life. Fittingly, it was the object of his last recorded observation, on 19 March 1885.

S Ori is a star of some interest, and the recent paper by Merchan Benitez and Jurado Vargas shows that it has an abnormal range of period, decreasing from 445 days to 397 days over a period of 16 years, and then increasing again. The extreme observed range of brightness is between V = 8 and 14. (See p. 211)

Star colours

Whilst the earlier observations of double stars and indeed any coloured stars found Webb using longhand notation, from the mid-1860s he began to record colours in shorthand. This took the form of a letter for the colour together with an index to denote the shade (ranging from 1 to 4); the greater the index the more subtle the shade. For instance, Y4 = pale yellow, Y2 = bold yellow, and combinations thereof – pale orange ruby (OR4), or good yellow with orange cast (O4Y3). Other colours included L = lilac, P = purple, and G = green. Webb was not drawn to the extended colour scheme of Smyth, who had listed a large number of hues – including fifteen shades of white! Smyth later re-evaluated this scheme.[13]

In estimating colour, Struve was much more conservative, but even he had to occasionally concoct a hue to represent what he saw, the most famous of which was 'olivaceasubrubicunda' – pinkish olive. Webb, in a burst of humour rare in the observing notebooks, gave a list of other words which had even more syllables! The red stars were favoured with descriptions such as carmine, ruby, garnet, scarlet and crimson.

Acknowledgements

The data in this chapter are from the WDS, maintained by the United States Naval Observatory. I am indebted to Dr Brian Mason for supplying the references to the first measures of the Webb double stars. I am also indebted to Dr David Dewhirst for supplying some very useful references to the SDUK.

Notes and references

1 W.H. Smyth, *A Cycle of Celestial Objects*, 2 vols. (J.W. Parker, London, 1844).
2 F.G.W.Struve, *Stellarum Duplicium et Multiplicium Mensurae Micrometricae* (Petropolis, 1837).
3 Alexander Jamieson, *Star Atlas* (London, 1822).
4 John Flamsteed, *Historiae Coelestis Britannicae* (published by the author, London, 1725).
5 R.A. Proctor, *A Star Atlas for the library, the School, and the Observatory* (Longmans, Green and Co., London, 1870).
6 Francis Baily, *Catalogue of Stars for the British Association for the Advancement of Science* (Taylor, London, 1845).
7 S.W. Burnham, *English Mechanic*, no. 426 (23 May 1873), 247.
8 S.W. Burnham, 'A General Catalogue of 1,290 Double Stars', *Publications of the Yerkes Observatory*, **1** (1900).
9 *Astronomische Nachtrichten*, **67**, 1591 (1866), 97.
10 *Transactions of the Royal Irish Academy*, **26** (1876), 249.
11 T.W. Webb, *Monthly Notices of the Royal Astronomical Society*, **50** (1879), 90.
12 *Monthly Notices of the Royal Astronomical Society*, **31** (1870), 84.
13 D.F. Malin and P.G. Murdin, *Colours of the Stars* (Cambridge University Press, 1984).

Chapter 15

Celestial objects for common readers

Webb as a populariser of science

Bernard Lightman

It is very pleasant to think that one may have been of some little use to a younger brother in a study in which, after all, we are all but children.[1]

In September 1864 Thomas Webb presented a paper 'On the Invisible Part of the Moon's Surface' at the meeting of the British Association for the Advancement of Science held at Bath. According to the British Association report, Webb had examined the theory that 'a different condition, both as to surface and atmosphere, might possibly obtain on its remoter side', and had asserted that with the evidence available it was not possible to form 'any satisfactory conclusion as to the condition of the invisible region'.[2] Another report on Webb's paper, offering a somewhat different account, appeared in the pages of *Punch*. Since its inception in 1841, the comic writers and artists of *Punch* had satirised scientific personalities and institutions, and rarely missed an opportunity to transform a serious scientific meeting into a farce. The British Association was a favourite target, as was the progressive ideology of science. *Punch* was especially fond of contrasting the grand ambition of scientists to uncover the mysteries of the Universe with the rather meagre results of their research.[3] Although the magazine was rarely considered as a source for accurate reports on the meetings of scientific societies, the short account of Webb's paper, entitled 'Advance in Astronomy', treated his work as a populariser of science perceptively. The title was deliberately ironic, for the whole point of the article was to cast doubt on whether Webb's paper advanced scientific knowledge at all. Webb unwittingly provided an ideal vehicle for *Punch* to expose the arrogance of astronomers to ridicule. 'For all that appears to the contrary,' *Punch* declared, 'that side of the Moon may be made of "green cheese".'[4] The critic in *Punch* found in Webb a like-minded comrade, for despite Webb's enthusiasm for exploring the heavens, his primary message hinged on the notion that the chief purpose of astronomy revolved around making the common reader more aware of their ignorance of a mysterious, divine creation.

Webb's activity as a populariser of science took place during a period when there was a veritable explosion of popular science writing. But what did Webb have to offer the rapidly growing reading audience in Britain in comparison to other popularisers of astronomy of the latter half of the nineteenth century, such as Richard Proctor, Agnes Clerke and Robert Ball? Webb's influential role as a populariser of science revolved around his emphasis on observational astronomy geared towards the amateur. Whereas other popularisers of astronomy made their reputations by exploring the exciting revelations of new scientific instruments or by spelling out the social or political implications of new discoveries, Webb maintained the traditional focus on the telescope, rarely strayed beyond the consensus of the astronomical community, and adhered to communicating the results of careful observation. Like other popularisers of astronomy, his view of the heavens was framed by religious concepts drawn from a theology of nature; but his religious beliefs led him to emphasise, as the article in *Punch* implied, that astronomers become more cognisant of the need for humility in the face of the divine power revealed – and concealed – by their telescopes.

In *The Victorian Amateur Astronomer*, Allan Chapman portrays Webb as a man who started a movement. Webb, the author asserts, was the father of modern amateur astronomy as 'a pursuit for serious observers whose principal motivation was pleasure, fascination, or the glory of God, as opposed to fundamental research'.[5] But if Webb founded this movement it was through his activities as a populariser of science. Webb came upon the scene when there was a renewed interest in astronomy at the same time that there was a growing market for popular science. In 1883 he pointed out to his readers that they were living in a period when 'a new and unprecedented interest has been awakened in astronomical pursuits' and 'competent instruments' have been placed 'within the reach of all classes'.[6] Cheap books and periodicals on popular science were also accessible for the first time from the middle of the century, as the growth of an educated middle class, combined with the invention of cheaper printing technologies, led to the growth of an unprecedented mass market that provided new opportunities for careers in lecturing, journalism and writing.

Like other popularisers of science, Webb was involved in a number of different types of activity, including public lecturing, contributing numerous articles to various periodicals, and composing books aimed at the general reader. As a public speaker Webb gave lectures at Cheltenham Ladies' College and to workers. In 1856, in the second of a series of free lectures for working men, Webb spoke on astronomical subjects at Cheltenham town hall. According to the report in the *Cheltenham Examiner* the audience consisted of 'about 600 of the class for whom the lectures are specially designed, and from about 200 to 300 of the upper class, amongst whom we are happy to observe a considerable number of ladies'.[7] Webb's involvement with Cheltenham Ladies' College seems to have lasted from at least the 1860s to the 1880s. In 1866 he was preparing twelve lectures on heat, electricity and magnetism for the College, and though he did not feel entirely qualified to lecture on new discoveries in these fields, he saw himself as 'providentially called' to undertake this duty.[8] Only a few years before his death he was still struggling with scientific theories in his lectures that went well beyond his expertise in astronomy. In 1881 he wrote to A.C. Ranyard: 'We hope to take flight to Cheltenham where I shall lecture a bevy of young girls a lot of stuff that I don't very well understand myself about luminiferous ether and such monstrosities.'[9]

Webb was a prolific contributor to popular science journals, as were many of his popularising colleagues. The *Intellectual Observer*, *Popular Science Review*, *Nature*, *English Mechanic* and *Knowledge* were among the periodicals to which he made substantial contributions. The vast majority of his articles deal with astronomical themes. The *Intellectual Observer* and *Popular Science Review* were founded in the same decade, and were part of the flourishing of popular science journalism in the 1860s which aimed at a middle-class readership.[10] Webb was in on the *Intellectual Observer* – a monthly journal – from the beginning in 1862 until the last issue in 1868. Virtually every month an article by Webb appeared dealing with the planets and occultations visible for viewing at that particular time of year, and he also wrote about double stars, the Moon, Uranus, comets, clusters and nebulae, solar observation, and telescopes. Although he contributed an occasional article to *The Student* – the successor to the *Intellectual Observer* – he no longer wrote monthly articles. Webb contributed much less to the *Popular Science Review*, although it survived from 1862 to 1881 – far longer than the *Intellectual Observer*. In the early 1870s a handful of articles deal with such topics as observations of Jupiter and how to sketch the Moon.

Beginning in the early 1870s Webb also became a frequent contributor to *Nature* – the weekly journal founded by his friend Norman J. Lockyer in 1869 to popularise the views of professional scientists.[11] Although there are some years when nothing from Webb appeared, more often an average of two to three of his pieces graced the pages of

Nature on an annual basis, until his death. Whether it was a letter to the editor, a review of a new book on astronomy, or articles reporting on his observations of some celestial phenomenon, Lockyer welcomed Webb's contributions. Webb's ability to move with ease between diverse scientific circles is demonstrated by his involvement with the *English Mechanic* and *Knowledge*. The *English Mechanic* – a cheap mass-circulation science journal run cooperatively with its largely working-class readers – was founded in 1865. It was a rival of *Nature*, and in 1870 boasted a circulation larger than all the other English scientific publications put together.[12] During the 1870s Webb often participated in its extensive correspondence columns, and also wrote articles on such topics as telescopes and other astronomical subjects. In the 1880s he also contributed a handful of articles on the canals of Mars, comets, and lunar delineation to the pages of *Knowledge* (an upmarket version of *English Mechanic*), founded by Richard Proctor in 1881. Due to Proctor's feud with Lockyer, his harsh criticisms of scientific societies, and his vision of an egalitarian scientific community, his list of contributors rarely overlapped with *Nature*'s.[13] As a contributor to *Nature*, *Knowledge*, *English Mechanic*, *Intellectual Observer* and *Popular Science Review*, Webb was highly unusual. Diverse popular science journals representing the scientific elite, the middle-class audience, or even working-class readers, all found room for the products of his prolific pen.

Webb also participated in a third area of popular science activity: writing books for the general reading public, such as *Celestial Objects for Common Telescopes* (1859), *Optics Without Mathematics* (1883), and *The Sun: A Familiar Description of His Phaenomena* (1885). In *Optics Without Mathematics* Webb explains the basics behind the transmission, reflection, refraction, dispersion and absorption of light, including a discussion of the more remarkable optical instruments. *The Sun* – an examination of solar astronomy for beginners – covers such topics as parallax, the distance and size of the Sun, sunspots, eclipses, and the role of the spectroscope in helping scientists to learn about the physical constitution of the Sun. But *Celestial Objects for Common Telescopes* was by far the most important and the most successful of all his books. Although telescopes had become less expensive, amateur astronomers were still confronted by a serious obstacle: their own inexperience. In the introduction, Webb declared that the purpose of *Celestial Objects* was to 'furnish the possessors of ordinary telescopes with plain directions for their use, and a list of objects for their advantageous employment'. By telling the serious amateur what to look for and how to look for it, Webb was filling a gap in the astronomical literature of the day. Although materials existed to guide the amateur, 'some of them are difficult of access', Webb declared, 'some, not easy of interpretation, some, fragmentary and incomplete'. For the more advanced observer he recommended W.H. Smyth's *A Cycle of Celestial Objects* (1844).[14]

Webb actually imitated the structure in Smyth's *Cycle*, beginning with telescopes, observing practices, and helpful hints, before moving on to a detailed description of planetary and stellar objects which the amateur could observe with an affordable telescope of low power. In Part I – 'The Instrument and the Observer' – he discusses what he means by a 'common telescope':

> By 'common telescopes' are here intended such as are most frequently met with in private hands; achromatics with apertures of 3 to 5 inches; or reflectors of somewhat larger diameter, but, in consequence of the loss of light in reflection, not greater brightness.

He goes on to give suggestions on how to distinguish good from faulty instruments, how to set up, operate and care for the telescope, and how to record observations. In Part II – 'The Solar System' – he discusses the Sun, Mercury, Venus, the Moon, Mars, Jupiter, Saturn, Uranus, Neptune, comets and meteors. In each case he draws attention to the distinctive features of each object, and to the history of the observations and theories of famous astronomers from the past. Part III – 'The Starry Heavens' – focuses on double stars, clusters and nebulae, and includes detailed descriptive catalogues.[15]

Celestial Objects was well received during Webb's lifetime, as each successive revised and enlarged edition was greeted by positive reviews across the full range of popular science journals. The second edition, published in 1868, was acclaimed by the *Popular Science Review* for its clarity. 'Indispensable to the beginner,' the reviewer asserted, it 'will develop many young astronomers, who by careful attention to its pages may immortalise themselves by discoveries in a department of science which is as fascinating to the ordinary thoughtful man as it is sublime to the philosopher.'[16] In 1873, when the third edition was published, the *English Mechanic* referred to *Celestial Objects* as a 'well-known and popular book' which was familiar to all readers who owned telescopes. 'We need not tell those who have this volume that it is worth buying,' the reviewer insisted, 'but those who have not yet made its acquaintance have an excellent opportunity afforded them by the issue of the new edition.'[17] The *Popular Science Review* reviewed the 'excellent little volume' again, praising the new material and remarking that it was so 'well-known that it is only necessary to say that a third edition has appeared to make the student purchase it'.[18] *Nature* was more subdued in its praise. Although *Celestial Objects* was referred to as a 'most useful adjunct' to the common telescope, and though Webb's advice on the use of telescopes and the mode of observation was deemed to be sound, the reviewer aired several criticisms. The detailed description of features of members of the Solar System might lead possessors of small telescopes to expect to see too much; the contrast of light and shade in some of the illustrations misrepresented what was seen through the telescope; and too much of the book was taken up with double stars and nebula, which would lead to mere stargazing. In keeping with the *raison d'être* of *Nature* as a spokesman for professional scientists, the reviewer wanted the book to teach amateurs 'how to make their observations of real use and not a mere pastime'.[19]

In 1881, when the fourth edition appeared, *Nature* was more favourable. 'That his volume will maintain its popularity amongst amateur astronomers is not to be doubted,' the reviewer declared, 'and we must add that it well deserves to do so.'[20] *The Observatory* refrained from discussing the general scope and design of *Celestial Objects*, as it was so well known and appreciated. After outlining the changes and additions to this new edition, the reviewer affirmed that the book was 'now more than ever an indispensable manual to all who possess a telescope and who wish to use it'.[21] Immediately after Webb's death in 1885, several astronomers, looking back on his accomplishments, treated his *Celestial Objects* as his crowning achievement. In *Nature*, G.F. Chambers mourned the loss of one of English astronomy's 'most assiduous and accomplished votaries'. Although Chambers acknowledged the significance of Webb's essays in scientific magazines, singling out his contributions to the *Intellectual Observer*, 'it was by his *Celestial Objects for Common Telescopes* that he became chiefly known in the astronomical world'.[22]

In the pages of his journal *Knowledge*, Richard Proctor agreed with Chambers' evaluation. No-one during the present generation, Proctor asserted, 'had done so much to popularise observational astronomy as the late vicar of Hardwick, whose *Celestial Objects for Common Telescopes* is employed as a handbook in nearly every observatory in the kingdom'. Although Webb was a prolific writer in various periodicals on astronomical subjects, *Celestial Objects* was 'the work with which his name is most widely associated'.[23] Webb's popular book outlived him. The Revd T.H.E.C. Espin revised two more editions – the sixth appearing in 1917.

Although Webb's *Celestial Objects* went through six editions, the small print-run for each edition has to be taken into account in gauging to what extent it can be considered a popular science best-seller. For the first edition in 1859, Longman published 1,000 copies. Of these, a little over half sold during the first year, while the rest were purchased by 1865.[24] Compared to a sold-out first edition of 1,500 copies of Darwin's *Origin of Species*, published in the same year, the sales of *Celestial Objects* do not appear to be particularly impressive. Moreover, the sales of both *Celestial Objects* and the *Origin of Species* were dwarfed by the 100,000 copies of John George Wood's *Common Objects of the Country* (1858) sold in the very first week![25] Wood – a populariser of natural history – had the advantage of being published in a Routledge series of shilling handbooks. By contemporary standards, then, *Celestial Objects* was a modest best-seller at first. The second edition run of 1,000 copies sold more quickly, so for the third edition of 1873 Longman produced 1,500 copies, nearly all of which had been sold by 1877.[26] Longman responded by publishing 3,000 copies of the fourth edition in 1881, and promptly sold 968 copies in the first year.[27]

By the time of Webb's death four years later, Longman had sold nearly two-thirds of the fourth edition, and approximately 5,500 copies in total.[28] Although sales increased steadily after the first edition was published, Webb's *Celestial Objects* did not have the broad appeal of the works of other popularisers of science. But since the book was 'designed for so limited a class as that comprised by working astronomers' – as the reviewer in *The Observatory* pointed out – it was certainly a great success within the astronomical community.[29]

Webb's role as a populariser of science is best understood by comparing him to others engaged in similar activities. In the second half of the nineteenth century it was not unusual for a Christian clergyman to be involved in popularising science. Webb was joined by Ebenezer Brewer – one of Jarrold and Sons' most successful authors, whose *A Guide to Scientific Knowledge of Things Familiar* (1847) was in its thirty-second edition by 1874. The Revd Charles Alexander Johns, whose *Flowers of the Field* (1853) sold well into the twentieth century, published widely on birds and botany. Johns encouraged a passion for botany in the influential broad churchman Charles Kingsley, his former student at Helston grammar school. Kingsley's *Water Babies* (1863) was only one of his many efforts to popularise science. The Revd George Henslow published a series of popular works on botany in the last three decades of the nineteenth century, while during the 1890s the Revd Henry Neville Hutchinson wrote about dinosaurs and geology.

Webb was part of that group of popularisers who combined their popular science activities with another job. He did not, like John George Wood, retire from regular clerical work to pursue a career as a populariser of science. He did not build a career as a populariser in order to make a living. Whereas Proctor was forced into becoming a populariser when an investment failed, Webb already had a 'living' and was a comfortably-off landowner from a well-connected family. Since Webb did not rely on his popular science work to provide a steady income, he was not as astoundingly prolific as some of his fellow popularisers. He did not lecture in public as often or in as many varied venues as Ball or Proctor. Proctor and Wood were dependent for their livelihood on their popularising activities, so they organised extensive lecture tours in England and abroad. Webb's contributions to periodicals were primarily to the popular science journals, unlike Clerke, Proctor and Ball, who also wrote for the general periodical press.[30] All in all, Webb produced far fewer pages of popular science writing and reached a more narrow audience than did Clerke and Proctor.

Webb also differed markedly from Clerke and Proctor in the way that he approached popular science writing. He and Proctor were friends. In addition to contributing articles to Proctor's journal *Knowledge*, Webb collaborated with him on a project directly connected to *Celestial Objects*. Proctor's *A New Star Atlas for the Library, the School and the Observatory* (1872) was intended to be a companion to Webb's *Celestial Objects*. Many of the objects dealt with in Webb's work were shown in Proctor's *New Star Atlas*. In the preface to his book, Proctor thanked Webb 'for the careful revision of the Atlas, so far as the objects to be included in his treatise are concerned'.[31] Nevertheless, Proctor and Webb adopted very different writing styles in their popular science works. Proctor attracted his readers by presenting controversial theories with power and imagination, and his fame as a populariser of astronomy rested on his vigorous defence of pluralism in his *Other Worlds Than Ours* (1870). Webb, however, was far more cautious. In his *Celestial Objects* he was willing to admit that Mars could be imagined to be habitable by man 'without much extravagance', but although initially receptive to Proctor's system of nomenclature for Mars, he was later unwilling to accept it as confirmed.[32] Yet, embedded in that nomenclature, which distinguished between land and water on Mars, was Proctor's case for the possibility of life on Mars.[33] In an article on Mars in *Nature* in 1874, Webb acknowledged that something corresponding to the outlines of lands and oceans had been more definitely recognised by astronomers over the past century, but he insisted that this view was not universally received.[34] Unlike Proctor, Webb's work did not focus on the issue of extraterrestrial life.

Webb also diverged from those popularisers of astronomy who emphasised the wondrous revelations of new scientific instruments. Proctor was particularly fond of the spectroscope, taking advantage of William Huggins' ground-breaking work in the 1860s, which gathered information on the chemical composition of celestial objects through a comparison of their spectra with terrestrial spectra. Proctor believed that the spectroscope was effecting a revolution in astronomy, and relied on it to provide evidence for his pluralist beliefs.[35] In contrast, Webb did not privilege the spectroscope. In *Celestial Objects* he referred to the 'revelations of the spectroscope' concerning the existence of twenty-three terrestrial elements in the Sun as being 'among the most surprising of modern astronomical discoveries'; but they 'still leave much to be explained by future investigation; much that we can never reasonably hope to explain'.[36] In an essay on 'The Great Nebula in Andromeda', Webb discussed the failure of the telescope to resolve disagreement among astronomers, and turned to spectroscopic results for help. He described the spectroscope as 'an instrument as superior in analytical as it is inferior in optical power'. Although the spectroscope yielded new information, it did not amount to a decisive answer to the most pressing questions.[37]

Nor did Webb become infatuated with the camera – unlike Agnes Clerke, whose activity as a populariser of astronomy began in the same decade that marked the death of both Webb and Proctor. From the middle of the 1880s to the middle of the first decade of the twentieth century, she brought the stunning results of astrophotography before the eyes of the public. Clerke believed that photography was the key to present and future advances in astronomical science. She once asserted that 'the whole future of astronomy has indeed come to depend on the validity of photographic evidence'.[38] Webb de-emphasised the potential of the camera in astronomy, and pointed to its limitations: 'There is a sharpness and decidedness in telescopic vision,' he argued in 1873, 'which we miss in photographic portraits.' The camera therefore failed to exhibit 'the true character, of the minutest features'.[39] Ten years later, after the quality of astronomical photographs had improved due to the introduction of the gelatine dry plate in 1875, Webb still refused to privilege the camera. Although photography 'misses nothing' and catches information which may be missed by an observer with 'fallible' vision, photographic appliances were not readily available, and the eye was capable of laying hold of the true form and character of an object in cases when the camera was rendered useless by atmospheric agitation. Webb granted that 'for the general mapping of an extensive region, and laying down its form and boundaries in due proportion, photography is by far the best suited; but there may always be room for filling-in detail by the eye'.[40]

Webb's approach to popularising astronomy had been set before the 'new astronomy', based on the spectroscope and camera, arrived on the scene in the last third of the century. He relied on a more traditional emphasis on telescopic observation and drawing, and saw his role as guiding his readers on how to operate a telescope effectively and how to draw what was seen through a telescope as accurately as possible. A true appreciation of astronomy, he believed, could only be obtained through first-hand participation. In the introduction to *Celestial Objects* he affirmed that:

> ... none but an eye witness of the wonder and glory of the heavens can thoroughly understand how much they lose by description, or how inadequate an idea of them can be gathered in the usual mode, from books and lectures.[41]

Photographs or analyses of spectra were no replacement for an immediate encounter with the skies.

Webb's writing style was in keeping with his goal of encouraging readers to engage directly with nature. Proctor contended that 'the gentleness and simplicity of his life were reflected in his writings', but Webb used simple language in order to make his work accessible to beginners – the hallmark of the writing of many successful popularisers of science.[42] The style of Webb's entries in his observing books reminds Chapman of the language used by naturalists when seeing a beautiful or rare flower, and he has referred to the culture of amateur science wherein others spoke of the flowers of the heavens as objects of sensory and spiritual delight;[43] indeed, Proctor wrote a book titled *Flowers of the Sky* (1879). What Chapman noticed about Webb's observing books can be extended to all of his popular science writings. Both the narrative style of popular science and the culture of amateur science were powerfully influenced by the traditions of natural history.

Popularisers of science offered accounts of nature – embedded in what scholars have referred to as a 'narrative of natural history' – that were diverting, highly descriptive, full of anecdotes, and non-theoretical. They focused on the singular and the extraordinary, on the unique beauty of nature, and they often communicated knowledge through telling a story which made the reader feel as though they were accompanying the writer on an adventurous quest for knowledge. Many popularisers also perpetuated a redefined and updated form of natural theology – an important theme in previous natural history. The features of this 'narrative of natural history' allowed the popular science writer to make science accessible to a popular audience.[44] Webb's popular science work shared many of these characteristics. He told anecdotes about astronomers of the past and about his own encounters with the celestial object under discussion; he emphasised detailed description of what was observed through the telescope; and he often attenuated the fantastic and the exotic. Webb appealed to the aesthetic sensibilities of his readers. Moreover, he attempted to engage their religious sensibilities – their sense of divine wonder.

To Webb, the primary purpose of studying the skies was to bring the astronomer closer to God.[45] In this sense, Webb's work as populariser was almost an extension of his clerical duties. In the introduction to his *Celestial Objects* he insisted that 'to do justice to this noble science' of astronomy meant appreciating firsthand 'the magnificent testimony which it bears to the eternal Power and Godhead of Him' who made the heavens.[46] Webb insisted that, though every science was 'full of the traces of the Creator', astronomy was the 'chief declaration of His Glory, the especial showing of His handiwork'.[47] He was not alone in sounding religious themes in his popular science works. Many other popularisers treated the purpose and order in nature in religious terms.[48] In the second half of the nineteenth century – when professional scientists such as T.H. Huxley worked to secularise science, to undermine natural theology by championing evolutionary theory, and to reduce the power of the Anglican clergy within British scientific institutions and societies – popularisers of science offered a revised and updated natural theology. The heavy-handed references to God's wisdom, goodness and power in classical natural theology, with its stress on specific instances of design and its use of scriptural quotations, was given up and replaced by a more subtle approach. The emphasis on design was more broadly conceived and located in the natural order or the laws of nature. Discussion of the divine meaning of nature was limited chiefly to introductions and conclusions, providing a sophisticated religious framework informing the entire work which was drawn on only occasionally in the text.[49] Professional astronomers such as Robert Ball – who were heavily involved in popular science activities – rejected any significant inclusion of religious themes. In his *The Story of the Heavens* (1885), the main *leitmotif* was the creativity of the human intellect in disclosing the wonder of the heavens, not the creative power of the deity.[50]

But Ball – whom Chapman refers to as the greatest of all the late Victorian and Edwardian popularisers of astronomy – did not find the religious framework in Webb's *Celestial Objects* so intrusive that it undermined the value of the book for observational astronomers. To Ball it was a 'charming handbook' which would 'greatly increase the interest' of the astronomer's work.[51] Webb's commitment to a more subtle approach to natural theology is evident throughout his popular science works. The beauty and order of the heavens draws from him the occasional comment on the deeper meaning of things. Mars, Webb declared, may be arranged in a manner corresponding to Earth 'by the Great Creator as the seat of life and intelligence'.[52] The only definite result of describing the many changes of form and colour in Jupiter was 'to deepen our sense of the wonders of creation, and our reverence for its First Great Cause'.[53] Similarly, Webb affirmed that the grand effect of occultations of planets or stars by the Moon 'may well convey a deep impression of the omnipotent Power and consummate Wisdom which orders its undeviating course'.[54] Even comets provided the occasion for a hint as to the religious significance of celestial objects. There was no need to fear that the Earth would collide with a comet, since 'every orbit is planned out, without the possibility of deviation, by infinite wisdom and paternal goodness'.[55]

In the opening of Part III of *Celestial Objects* – 'The Starry Heavens' – Webb asserts that the instances of order and beauty in the Solar System are alone sufficient to bring home to the astronomer the presence of the divine. 'If the Solar System had comprised in itself the whole material creation,' Webb stated, 'it would alone have abundantly sufficed to declare the glory of GOD, and in our brief review of its greatness and its wonder we have seen enough to awaken the most impressive thoughts of His power and wisdom.' But our Solar System is 'but as a single drop in the ocean', and Webb promised that the third section of his book would deal with thousands of systems which contain 'more amazing regions, and fresh scenes will open upon us of inexpressible and awful grandeur'. Yet, all of them are 'bound together by the same universal law which keeps the pebble in its place upon the surface of the Earth, and guides the falling drop of the shower, or the mist of the cataract'.[56] The catalogue of double stars, clusters and nebulae was the key to viewing 'this great display of the glory of the Creator'.[57]

Religious themes surfaced from time to time in Webb's other books, written near the end of his life. In the opening pages of *Optics Without Mathematics*, he reminded his readers that light was one of the 'very best and chiefest gifts of Him who is pleased to describe Himself under the name of Light'. Throughout the book he stresses the wonderful properties of light, and then, in the conclusion, links them to God's 'wonderful works'. The purpose of the book, he declared, had been to teach the reader enough about light to awaken 'a greater interest than we had before, in wonders that lie on every side of us, but pass unnoticed because we see them every day'.[58] In *The Sun*, Webb's goal was to overwhelm the reader with the magnitude of the heavens in terms that even a child would understand. The distance between the Earth and the Sun was so great that 'walking 4 miles an hour and 10 hours every day, a man would travel as far as the Sun in 6,300 years'. If we took an express train travelling at 60 mph we would die before our journey of 175 years had ended. Later, in a discussion of the size of the Sun, he again tried to boggle the reader's mind with numbers. The Sun's 866,400-mile diameter draws from him the remark: 'What a stupendous ball of fire!'; and to emphasise his point he observed that '109 of our Earths touching one another would hardly cross the Sun from side to side', and an express train would take more than seven years to make a round trip. Webb concluded his book by asserting that the 'magnificent Sun ... a mighty exhibition of Creative Power and Wisdom, is but one among the countless host of heaven; is no other than a star!'.[59]

Webb's reflections on the greatness of the Creator were often coupled with discussions of the ignorance of astronomers. This is a theme which runs throughout his work, from the appearance of *Celestial Objects* in 1859 to his last popular science essays in *Nature* shortly before his death. In the introduction to *Celestial Objects*, Webb pointed to the value of astronomy as a leisure activity which also led to 'the most impressive thoughts of the littleness of man, and of the unspeakable greatness and glory of the Creator',[60] and in a number of his articles in the *Intellectual Observer* in the latter half of the 1860s he picked up on the same theme. An examination of the disagreement between astronomers as to the form of the shadows which Saturn and its ring mutually cast upon each other led him to note how ironic it was that 'the mystery of the subject has increased under closer, more powerful, and more extended scrutiny'. To those disappointed that no progress had been made in this area of study, Webb reminded them that research into the heavens brought its own compensation: an appreciation of God's great works.[61] The impossibility of determining whether or not a lunar atmosphere existed was the cue for him to make a plea for more research, which he 'commended to those who love to trace the footsteps of the Maker of all things in the manifold exercise of His creative power'.[62]

In his articles in *Nature* in the 1870s and early 1880s, Webb continued to explore the theme of human ignorance in the face of the divine wonders of the heavens. Although the unprecedented multiplication of telescopes in his time meant that Jupiter had never been 'subjected to such an extended scrutiny as the present', he impressed upon the reader the disagreement between observers. Due to the differences in the instruments used, in the sharpness of the eyes of astronomers, and in their degree of experience, agreement on the appearance of Jupiter was hard to obtain.[63] Astronomers fared no better when it came to the other planets. The best observers could not reach agreement as to the features of Mars.[64] Despite advances in optical power, little progress had been made in recent years in penetrating Saturn's mystery. 'What material progress have we to boast of?' Webb asked. 'What further light have the same instruments, or others of greater power, thrown on the minute subdivisions of the rings, or the abnormal and inexplicable outlines of the shadow of the globe?' Since Saturn was unique, and as astronomers could find no analogy in human experience, they were confronted by their 'entire ignorance of the real nature of our subject'.[65]

Many other celestial objects were equally elusive. In 'The Theory of Sunspots', Webb again stressed the ignorance of astronomers, apparent in their inability to reach agreement as to the true nature of sunspots. The observer of the solar disc 'knows absolutely nothing as to what he is looking upon', Webb declared. The best astronomers offered no help. 'Shall we listen to Wilson,' Webb asked, 'or Herschel, or Kirchhoff, or Nasmyth, or Secchi, or Faye, or Zöllner, or Langley? More or less, they all disagree.' In light of the protracted discussion over this issue, observers could 'hardly bring to our telescope an unbiased eye or an impartial judgment'. Solar phenomena lent themselves to 'very dissimilar and even opposite interpretations'. If the telescope disappoints us, Webb asserted, then perhaps the spectroscope could resolve the problem. But here too Webb found only equivocal evidence which was sometimes very perplexing.[66]

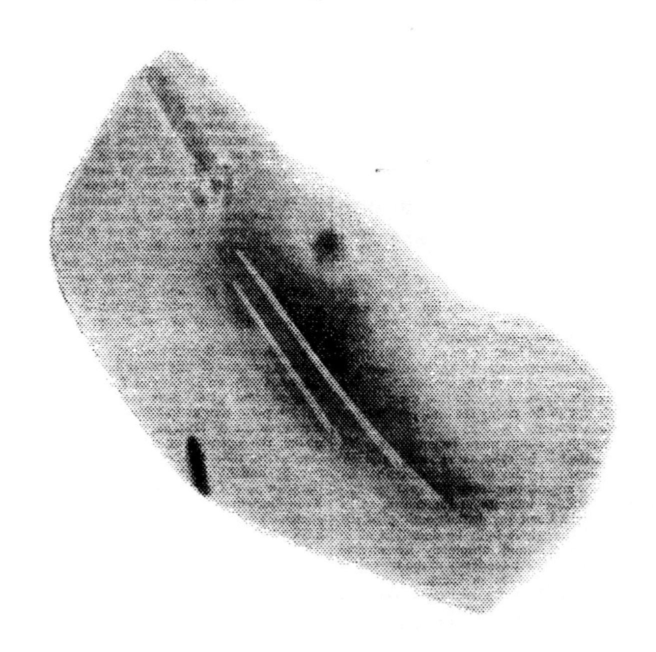

The Great Nebula in Orion, as drawn by Professor G.P. Bond in 1847. This is one of three different drawings of the nebula presented by Webb to demonstrate the depths of human ignorance. (T.W. Webb, 'The Great Nebula in Andromeda', *Nature*, **25** (9 February 1882), p. 341.)

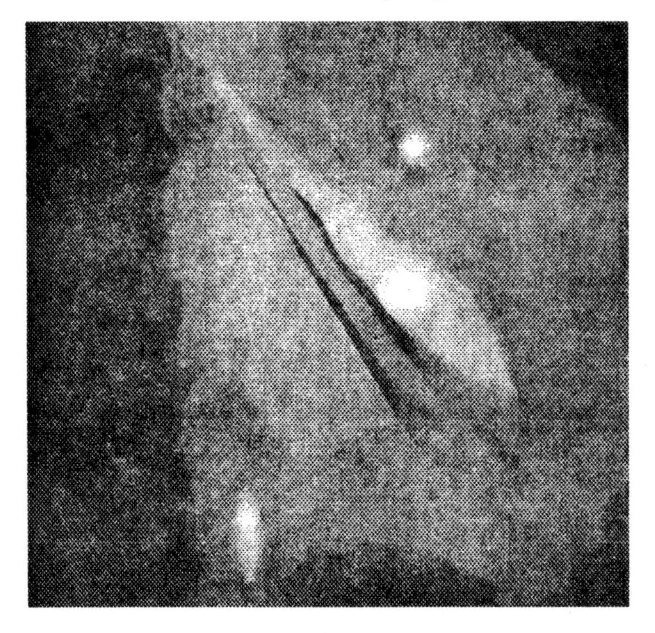

Trouvelot's depiction of the Great Nebula in Orion in 1874. (T.W. Webb, 'The Great Nebula in Andromeda', *Nature*, **25** (9 February 1882), p. 344.)

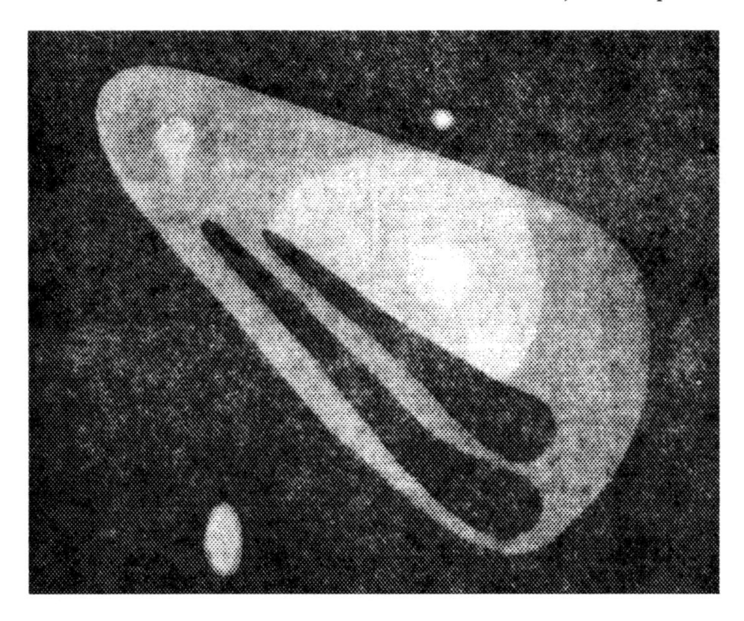

The Great Nebula in Orion, drawn by the Revd Jevon J. Muschamp Perry in 1881. (T.W. Webb, 'The Great Nebula in Andromeda', *Nature*, **25** (9 February 1882), p. 345.)

Unsurprisingly, he found astronomers to be even more unenlightened as to the nature of celestial objects outside the Solar System. In his article on 'The Great Nebula in Andromeda', this mysterious object became a symbol of human ignorance. Webb speculated that the Andromeda Nebula, despite its 'enormous magnitude', had been neglected by astronomers because it had 'hitherto resisted all inquiry'. After outlining the history of the nebula (at that time it was not known that it is a galaxy), he went on to discuss the disparity between observers. He examined three different drawings of the nebula made in 1847, 1874 and 1881 with the aid of telescopes of the greatest power, and then compared them to the results produced by astronomers working with smaller instruments. He considered the possibility that the discrepancies between these drawings were due to changes in the nebula over time, but could not rule out the possibility that the form of the nebula eluded human accuracy. The discrepancies in the drawings were 'illustrative of the uncertainty that hangs about such observations', and he concluded that the 'telescope has comparatively failed'. Even the spectroscope offered indecisive evidence as to the physical constitution of the nebula. As the largest body in the Universe, Webb described it as 'the greatest display as to magnitude of its incomprehensible Creator'. He ended the article with a profound admission of the limits of astronomy: 'And with these inquiries as to a mystery never in all probability to be penetrated by man, our imperfect remarks shall close.'[67]

Webb's emphasis on the relative ignorance of the astronomer and the glory of the divine heavens was actually quite provocative, particularly when it appeared in the pages of a journal such as *Nature*, which professional scientists like T.H. Huxley had hoped would become the chief organ for disseminating a secular vision of nature to a popular audience. Webb may not have intended to be provocative. In 1883 he confessed that it was not to his taste to sign his name to a review in *Nature*. 'I love to be more quiet,' he wrote to Ranyard, 'and I would not give it any time when it might be more likely to lead to what I so cordially dislike – paper-skirmishing.'[68] But although Webb attempted to evade controversy, his simple expression of awe and humility in the face of the heavens reflected his conservative, Anglican, and landowning background. Surely he did not attempt to use astronomical science explicitly to support a particular political position in the manner of a John Pringle Nichol, who pressed the nebular hypothesis into the service of political reform.[69] However, Webb's insistence on the ignorance of astronomers and the corresponding majesty of divine creation flew in the face of the positivistic inclinations of scientific naturalists. Huxley and his allies were interested in stressing the boundless capacity for knowledge to increase and in convincing the common reader that scientists had already discovered enough knowledge to qualify them, and not the Anglican clergy, as the proper cultural authorities for the modern, industrialised age.

Despite the growing power of scientific naturalism within Victorian science, Webb perceived his main role as populariser in religious terms. Even in the section on 'The Mode of Observation' near the beginning of *Celestial Objects*, he was reluctant to allow the reader to forget the religious importance of astronomy. After presenting a long list of tips on how to use the telescope and record observations, he concludes by reminding the 'young observer not to lose sight of the immediate relation between the wonderful and beautiful scenes which will be opened to his gaze, and the great Author of their existence'. He then draws an analogy between a beautiful painting, an ingenious piece of mechanism, and the exquisite heavens. Just as we credited artists with talent for the excellence of their works, Webb affirmed, and the designer of a clever machine with skill, we must not disconnect 'perfect creations' in the sky, 'so far transcending every imaginable work or art, from the remembrance of the Wisdom which devised them, and the Power which called them into being'. Telescopes were not to be used by common readers just as a source of 'mere amusement or curiosity'; rather, the 'right use of the Telescope' was to gain 'a more extensive knowledge of the works of the Almighty'.[70]

Like many other popularisers of science, Webb wanted to discipline the eyes of his common readers so that when they looked through their common telescopes they would see the hand of God behind the celestial objects in the skies.

Acknowledgements

The author wishes to thank his research assistant, Liza Piper, for supplying him with samples of the rich periodical literature by Webb; Peter Hingley, for information on the location of Webb's articles in journals; and Mark and Janet Robinson, for providing key selections from Webb's correspondence and other assorted materials on Webb's public lectures.

Notes and references

1 T.W. Webb, quoted by Arthur Mee in 'The Rev. T. W. Webb: In Memoriam', *Observational Astronomy* (Owen, Cardiff, 1893).
2 T.W. Webb, 'On the Invisible Part of the Moon's Surface', *Report of the Thirty-Fourth Meeting of the British Association for the Advancement of Science, held at Bath in September 1864* (John Murray, London, 1865), p. 9.
3 J.G. Paradis, 'Satire and Science in Victorian Culture', in B. Lightman (ed.), *Victorian Science in Context* (University of Chicago Press, 1997), pp. 143–75.
4 'Advance in Astronomy', *Punch*, **47** (1864), 143.
5 A. Chapman, *The Victorian Amateur Astronomer: Independent Astronomical Research in Britain 1820–1920* (Praxis–Wiley, Chichester, 1998), p. 225.
6 T.W. Webb, 'On Lunar Delineation', *Knowledge*, **4** (16 November 1883), 302.
7 'Free Lectures to Working Men', *Cheltenham Examiner* (20 February 1856), p. 4.
8 Webb to Ranyard, 29 January 1866.
9 Webb to Ranyard, 29 December 1881.
10 Ruth Barton, 'Just before *Nature*: the purposes of science and the purposes of popularisation in some English popular science journals of the 1860s', *Annals of Science*, **55** (1998), 1–33.
11 Webb and Lockyer struck up a close friendship after meeting at the Royal Astronomical Society. Webb helped Lockyer and his wife with their English translation of Edouard Guillemin's *Le Ciel*. See T. Mary Lockyer and Winifred Lockyer, *Life and Work of Sir Norman Lockyer* (Macmillan, London, 1928), pp. 22, 26.
12 R. MacLeod, *Public Science and Public Policy in Victorian England* (Variorum, Aldershot, 1996), 224.
13 B. Lightman, '"Knowledge" Confronts "Nature": Richard Proctor and Popular Science Periodicals', in Louise Henson *et al.* (eds.), *Culture and Science in the Nineteenth-Century Media* (Ashgate Publishing, Aldershot, 2004), pp.199–210.
14 T.W. Webb, *Celestial Objects for Common Telescopes*, second edition (1868), pp. vii–viii.
15 Chapman (n. 5), p. 225; Webb (n. 14), p. 1.
16 Anon., 'Star-Finding', *Popular Science Review*, **7** (1868), 301.
17 Anon., review of *Celestial Objects for Common Telescopes*, in *English Mechanic*, **17** (11 April 1873), 81.
18 Anon., 'A Companion to the Common Telescope', *Popular Science Review*, **12** (1873), 293–4.
19 G.M.S., 'Our Book Shelf: Celestial Objects for Common Telescopes,' *Nature*, **8** (10 July 1873), 199.
20 Anon., 'Celestial Objects for Common Telescopes', *Nature*, **25** (1 December 1881), 99.
21 Anon., 'Webb's *Celestial Objects*', *The Observatory*, **5** (1882), 11–13.
22 G.F. Chambers, 'The Rev. T. W. Webb', *Nature*, **32** (11 June 1885), 130.

23 [Richard Proctor], 'Editorial Gossip', *Knowledge*, **7** (5 June 1885), 481.
24 *Archives of the House of Longman, 1794–1914* (Cambridge, Chadwyck-Healey, 1978), A5, 599.
25 A. Desmond and J. Moore, *Darwin* (Michael Joseph, London, 1991), pp. 477–8; Richard Altick, *The English Common Reader* (University of Chicago Press, 1983), p. 389.
26 *Archives of the House of Longman, 1794–1914* (n. 24), A10, 399.
27 Ibid., A13, 435.
28 Ibid., A13, 436.
29 Anon. (n. 21), 11.
30 Clerke, Proctor and Ball all wrote for the *Contemporary Review*, Ball and Clerke contributed to the *Quarterly Review*, and Ball and Proctor had essays in the *Fortnightly Review*. In addition, Clerke's articles appeared in the *Dublin Review*, *Fraser's Magazine*, *Macmillan's Magazine* and, by far the most, in the *Edinburgh Review*. Pieces by Proctor were published in the *Cornhill Magazine*, *Temple Bar*, *Longman's Magazine* and the *Nineteenth Century*. All these were representative of the general periodical press.
31 Richard A. Proctor, *A New Star Atlas for the Library, the School and the Observatory* (Longmans, Green & Co., London, 1889), p. viii.
32 Webb (n. 14), p. 119. Webb reproduced Proctor's map of Mars (first published by Proctor in 1868) in the third (1873) edition of *Celestial Objects*, but dropped it from the fourth (1881) edition.
33 B. Lightman, 'The Visual Theology of Victorian Popularizers of Science', *Isis*, **91** (December 2000), 661–71.
34 T.W. Webb, 'Mars', *Nature*, **9** (12 February1874), 287.
35 Lightman (n. 33).
36 Webb (n. 14), p. 29.
37 T. W. Webb, 'The Great Nebula in Andromeda', *Nature*, **25** (9 February 1882), 343–4.
38 Agnes Clerke, *Problems in Astrophysics* (Adams and Charles Black, London, 1903), 5. See also Lightman (n. 33), 671–9; and B. Lightman, 'Constructing Victorian Heavens: Agnes Clerke and the "New Astronomy"', in Barbara T. Gates and Ann B. Shteir (eds.), *Natural Eloquence: Women Reinscribe Science* (University of Wisconsin Press, 1997), 61–75.
39 T.W. Webb, 'How to Sketch the Moon', *Popular Science Review*, **12** (1873), 235.
40 T.W. Webb, 'Our Lunar Delineation', *Knowledge*, **4** (16 November 1883), 302.
41 Webb (n. 14), p. vii.
42 [R.A. Proctor] (n. 23), p. 481.
43 Chapman (n. 5), p. 227.
44 B. Lightman, 'The Story of Nature: Victorian Popularizers and Scientific Narrative', *Victorian Review*, **25** (Winter 2000), 1–29.
45 See Chapter 7.
46 Webb (n. 41).
47 T.W. Webb, *The Sun: A Familiar Description of His Phaenomena* (Longman,

Green & Co., 1885), p. 2.

48 B. Lightman, 'The Voices of Nature: Popularizing Victorian Science', in B. Lightman (ed.) *Victorian Science in Context* (University of Chicago Press, 1997) pp. 187–211.

49 Lightman (n. 44).

50 Ibid., 20–4.

51 Sir Robert Ball, *The Story of the Heavens* (Cassell, London, 1905), p. 28.

52 T.W. Webb, 'Mars', *Nature*, **9** (12 February 1874), 287.

53 T.W. Webb, 'The Planets of the Season', *Nature*, **20** (23 October 1879), 606.

54 Webb (n. 14), p. 71.

55 T.W. Webb, 'The Earth in the Comet's Tail', *Intellectual Observer*, **1** (1862), 67.

56 Webb (n. 14), p. 168.

57 Webb (n. 14), p. 170.

58 T.W. Webb, *Optics Without Mathematics* (SPCK, London; E. & J. B. Young, New York, 1883), pp. 4, 121.

59 Webb (n. 47), pp. 26, 31, 78.

60 Webb (n. 14), p. x. When discussing Uranus and Neptune, Webb speculated that the advance of optical power in telescopes could prove that 'as far as the dominion of our own Sun is concerned, we have reached the boundary of our knowledge', rather than opening 'fresh marvels'. Ibid., p. 155.

61 T.W. Webb, 'The Planet Saturn', *Intellectual Observer*, **10** (October 1866), 201.

62 T.W. Webb, 'Gruithuisen's City in the Moon – Jupiter's Satellites – Occultations', *Intellectual Observer*, **12** (October 1867), 222.

63 T.W. Webb, 'The Planet Jupiter', *Nature*, **3** (30 March 1871), 430.

64 T.W. Webb, 'The Planets of the Season: Mars', *Nature*, **21** (1 January 1880), 213.

65 T.W. Webb, 'Saturn', *Nature*, **31** (26 March 1885), 485.

66 T.W. Webb, 'The Theory of Sunspots', *Nature*, **30** (15 May 1884), 59.

67 Webb (n. 37), pp. 341–5.

68 Webb to Ranyard, 2 February 1883.

69 S. Schaffer, 'The Nebular Hypothesis and the Science of Progress', in James R. Moore (ed.), *History, Humanity and Evolution: Essays for John C. Greene* (Cambridge University Press, 1989), pp. 131–64.

70 Webb (n. 14), p. 17.

Appendix 1

The Webb Society: its history and activities

Robert W. Argyle

> *I had at one time projected a survey of the wonders of the region [Cygnus] with a sweeping power; but want of leisure, an unsuitable mounting, and the astonishing profusion of magnificence, combined to render a task hopeless for me which, I trust, may be carried through by some future observer.*

The library of the British Astronomical Association occupies a small room on the top floor of the Royal Astronomical Society's premises in Burlington House, Piccadilly, London. The library is modest, but it was, and still is, a popular venue for amateur astronomers, both for the purposes of study and informal social meetings. During much of 1966 and 1967, Kenneth Glyn Jones was collecting material for a book on the Messier objects, to be called *Messier's Nebulae and Clusters*, and would often ascend the steep stairs to the BAA offices after the regular Wednesday meetings of the Association.

Kenneth Glyn Jones (1915–1995) was a man of many talents. He served with the RAF in the Far East during the Second World War, and in 1946 joined the staff of BOAC at Heathrow. He became a Senior Technical Instructor, at first specialising in astro-navigation and later turning to aircraft performance when inertial methods became standard. Many Boeing 747 and Concord pilots received their training in navigation at his hands. He retired from British Airways in 1974, but remained active. He was a skilled instrument maker and wood sculptor, and his house in Berkshire was filled with examples of his work. He built, among other things, an eclipse predictor, and a grandfather clock which shows phases of the Moon. His interest in historical

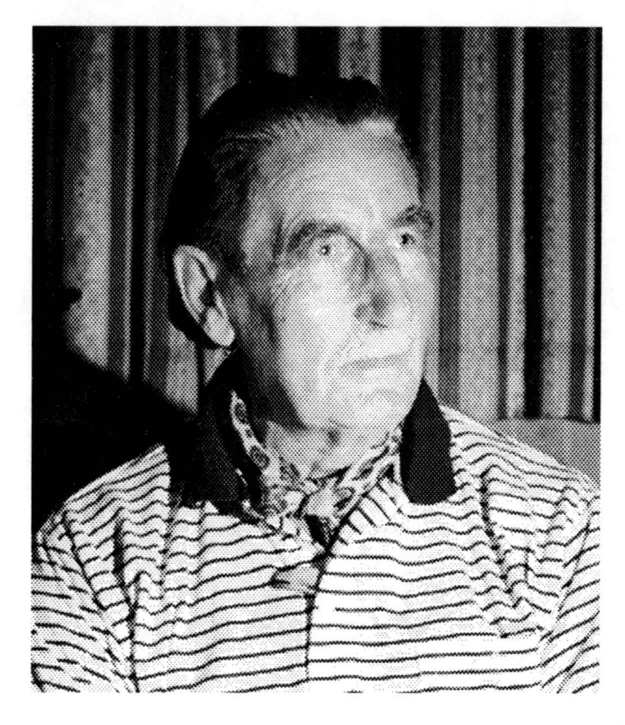

Kenneth Glyn Jones, co-founder of the Webb Society.

astronomy led him to write an eponymous nine-part series of papers, for the *Journal of the British Astronomical Association*, about the search for the nebulae, and these were later gathered together and published as a small book. He later went on to write *Messier's Nebulae and Clusters*, which is still the standard reference work on this catalogue of objects.

It was during one of Kenneth Glyn Jones' visits to the BAA library that he met James Muirden – a skilled telescope builder, and author of *Astronomy with Binoculars*. In 1964 Muirden had founded a magazine called *The Casual Astronomer*, for the purposes of propagating observations more quickly than had previously been possible. The BAA *Journal* was (and still is) issued bi-monthly, and following up observations of novae and comets usually required a telephone call or a postcard. At that time the International Astronomical Union had a telex service for comets, supernovae and novae, but the messages were sent as a stream of five-digit numbers, and had to be laboriously translated at the receiving end.

With Muirden as Editor, John Larard occupied the position of Secretary and assistant. A keen double-star and deep-sky observer, he took the opportunity to write a column called 'From the Night Sky', which first appeared in Vol. 2 No. 14 (June 1965) of *The Astronomer* (*TA*), as the magazine had now become. At that time the BAA had no Deep Sky Section (it did not form one until the early 1980s), and the Double Star Section had closed in 1912. Larard was a great admirer of T.W. Webb, and, encouraged by the success of the 'From the Night Sky' column and the interest it engendered, he suggested the formation of a society dedicated to deep-sky observing. The first idea was mooted in the pages of *TA*, with Muirden's enthusiastic encouragement, and the Webb Society – with Glyn Jones as President and Larard as Secretary and Treasurer – came into being on 12 June 1967. Membership was open to all interested in the deep sky and double stars, and the subscription was 10 shillings per year. Thanks to a classicist of John Larard's acquaintance, the Society also had a motto, *Caeli Scrutamur Plagas – We Sweep the Regions of the Heavens* – from the works of the poet Ennius. Two observing sections were formed at this time: Nebulae and Clusters (directed by Glyn Jones) and Double Stars (run by Larard).

For the first year the Society conducted its business through the pages of *TA*, and quickly reached a membership of forty, including amongst its number John Isles and the author (who are still active in the group today), and David Allen who, sadly, was to die prematurely in 1994. On the first anniversary of its formation in June 1968, the first issue of the *Quarterly Journal* appeared. It was typed on stencil, in A4 format, by John Larard, and its twenty-six pages included such articles as 'The Twentieth Century Webb' by James Muirden and 'Gaseous Nebulae in Orion' by David Allen.

By the second anniversary in 1969, membership had reached sixty-five, and the future looked bright. But then John Larard suffered a long spell in hospital, and the production of the *Quarterly Journal* and even the future of the Society became distinctly doubtful. Happily, John recovered, and production resumed on an even keel. In the summer of 1969 it was decided to hold an Annual General Meeting. The venue was the large garden of Kenneth and Brenda Glyn Jones, in Winkfield, just outside Windsor. The meeting duly went ahead on 5 September, the weather was benevolent, and sixteen members attended, including three from Newcastle-on-Tyne. Talks were given by David Allen, Jim Hysom and Ian Genner. David Allen described a recent trip to Minnesota to carry out some observations in the infrared. Although he was to go on to become one of the world's leading infrared observers, he always had time to both talk and write for the Society, and was always a very welcome speaker at annual meetings. In 1975 he went to work at the Anglo-Australian Observatory, and became the first permanent member of staff.

One of the aims of the Webb Society, from the very outset, was to produce a new observing handbook of deep-sky objects and double stars – not to replace Webb's *Celestial Objects*, but to complement it and to update the entries on each object. Rather than produce a modern equivalent of the second volume of *Celestial Objects* it was decided to tackle it in two instalments, beginning with a volume on double stars. This was authored mainly by Robert Argyle, and the first version, produced on a Gestetner by Eddie Moore, appeared in 1975. The 350 copies in the first run sold steadily, and a reprint of 300 copies, produced two years later, vanished in similar fashion.

Following the production of this volume, a small number of enthusiastic double-star observers began to send in series of micrometric observations. In the first Double Star Section *Circular* in 1979, most of the measures were made by the grating micrometer as a direct consequence of the device being featured in the *Handbook of Double Stars*. (This micrometer – which is easy to construct – can produce accurate results on bright wide pairs, but as it is an objective grating there is a heavy price to pay in loss of light in the eyepiece: 50 per cent in the usual design. At this time there were no commercially available filar micrometers, but it was a beginning.) The *Handbook* sold successfully, so it was decided to go ahead with a volume on nebulae and clusters. Co-authored by David Allen and Edmund Barker, this too sold well, and reflected the growing interest in deep-sky observing in the UK and abroad.

In 1977, at the seventh AGM held at the Royal Observatory, Greenwich, it was suggested that a small number of Webb Society members might be allowed to use the 28-inch refractor. Programmes of observation of double stars and planetary nebulae were drawn up, and a number of visits to Greenwich were made. Unfortunately, the reality did not live up to the promise, with the poor weather combining with difficulty of access to make the project rather impractical.

The next decade saw the number of members slowly rise, and the handbook series took off with a further four volumes: Globular and Open Clusters, Galaxies, Clusters of Galaxies, and Anonymous Galaxies. For all of them, professional astronomers of note were asked to contribute forewords – which they duly did. An agreement with an American publisher – Ridley Enslow, of New York – allowed the Society to concentrate on astronomy. In 1986, continuing demand for the *Handbook of Double Stars* saw a revised version produced by Enslow. Steven Hynes then took on the task of producing a volume on *The Southern Sky* – which was rather a risky enterprise, as it was heavily based on the author's observations during occasional holidays in the tropics! The last handbook (Volume 8), on variable stars, was written by John Isles, who by that time was living and working in the United States after a period in Cyprus.

With the series of handbooks finished, the Society was left with a large accumulation of observations, and it was decided to launch a new publication, *Observing Section Reports*, as a medium through which to present the regular observations from the Nebulae and Clusters, Southern Sky, Double Star and Galaxies Sections. By the time that the agreement with Enslow was concluded, some 12,000 handbooks had been sold to individuals and bookstores. The number sold through the various book clubs is not known, but it seems probable that it was equally as many.

In 1992 the *Quarterly Journal* seemed in need of a facelift. The articles tended to be rather academic, and it was felt that contributions with a lighter touch would be more appropriate. Steven Hynes suggested the introduction of a new publication, the *Deep Sky Observer*, which would fill this niche. Initially, the new publication ran alongside the *Quarterly Journal* and alternated every six months with the *Observing Section Reports*. By October 1996 it was decided to expand the *Deep Sky Observer* from A5 to A4 format to accommodate larger photographs and drawings. Although at this point the membership had topped 400, it did not lead to a commensurate increase in the amount of material for publication, and it was no longer viable to support eight issues per year of the various publications. With *Quarterly Journal* No. 111 (1997), the *Deep Sky Observer* and *Quarterly Journal* combined, and the *Observing Section Reports* ceased to be.

In recent years, the activity of the Society has expanded to embrace the publication of a number of observing guides which range from a monograph on non-existent clusters in the *Revised New General Catalogue* to a visual atlas of double stars and a beginner's guide to deep-sky observing. Modern technology has also allowed us to maintain links with the observers of the past. The widespread availability of high quality digital cameras has encouraged us to reproduce a number of rare historical volumes, not easily accessible to most observers. To date, a first edition of *Celestial Objects*, the 1847 catalogue of observations in South Africa by John Herschel, and the Birr observations of Lord Rosse are available on compact disc. To encourage observers to send in contributions from all over the world, a Southern Sky Section was formed and is run by Jenni Kay, a keen and able observer who lives in South Australia. The *Deep Sky Observer* continues to be the primary organ through which the Society members keep in touch, and recent issues have contained colour plates. The website – ably organised by Tim Walker – also reflects activities, and contains links to members' websites as well as details of impending meetings and the available range of Society publications (www.webbsociety.freeserve.co.uk).

The Double Star Section is now in its thirty-seventh year, and the number of *Circulars* has reached thirteen, with the publication now appearing annually due to the number of observations received. Measures are now made with filar micrometers, CCD cameras, and with eyepiece micrometers of various types such as the Celestron Micro Guide eyepiece and the ring micrometer. An increasing feature is the use of Internet-based software to derive parameters of double stars from on-line Schmidt plate archives. The data in these *Circulars* are regularly sent to the United States Naval Observatory, in Washington, for incorporation into the *Observations Catalogue*. To date, more than 24,000 mean measures have been published.

Since its inception the Webb Society has contributed significantly to the encouragement of deep-sky observing in the UK and elsewhere, through the medium of the *Quarterly Journal* and the observing handbooks. The availability of large telescopes, personal computers and CCD cameras has largely led to a more technical – it might almost be said, 'professional' – approach to observing, but the Society remains strictly amateur in the best sense of the word; and, of course, the officers all freely give their time. Whilst recognising the trend towards electronic imaging, the visual observer is still encouraged, and this is reflected in the papers that are published in the *Deep Sky Observer* and the talks that are arranged at the annual meetings.

Acknowledgements

This appendix leans heavily on the five papers about the early history of the Webb Society, written by Kenneth Glyn Jones, and published in the Society's *Quarterly Journal*, Nos. 68–72 (1987–88).

Appendix 2

Bibliography of Webb's published works

Jaspreet Gill and Bernard Lightman

How little the best books on any subject deserve the title of exhaustive – I have long noticed this in astronomy.

T.W. Webb, cited by Arthur Mee

Besides writing three books – *Celestial Objects for Common Telescopes* (1859, 1868, 1873, 1881, 1893, 1917 and 1962), *Optics Without Mathematics* (1883), and *The Sun: A Familiar Description of His Phaenomena* (1885) – Webb was an extremely prolific contributor to academic and popular science periodicals. The following bibliography – which is extensive, but not exhaustive – is based on bibliographies by Peter Hingley and Liza Piper, with the help of Katrina Sark and Mark Robinson.

1835

Hints to Observers of Halley's Comet, *The Analyst: A Quarterly Journal of Science, Literature, Natural History, and the Fine Arts*, **3** (1835), 134–8.

The Moon, *Fraser's Magazine for Town and Country* (December 1835).

1836

On the Influence of Comets, *The Analyst*, **3** (1836), 218–29.

1839

On Lunar Volcanoes, *Report of the Eighth Meeting of the British Association for the Advancement of Science*, Newcastle, August 1838, **7**, Part 2 (1839), 93.

1853

Observations of Luminous Meteors, from 1818 to 1850, extracted from old diaries of natural phaenomena, *Report of the Twenty-Second Meeting of the British Association for the Advancement of Science,* Belfast, September 1852 (1853), 178–89.

On the variability of the light of the star β Corvi, *Monthly Notices of the Royal Astronomical Society*, **13** (8 April 1853), 186.

On the Small Star near α Canis Minoris and on the suspected variable star β Corvi, *Monthly Notices of the Royal Astronomical Society*, **13** (10 June 1853), 271.

1854

On the Zodiacal Light, *Monthly Notices of the Royal Astronomical Society*, **14** (13 January 1854), 83–5.

Note on the Zodiacal Light, the Companion of Procyon and Sirius, *Monthly Notices of the Royal Astronomical Society*, **14** (12 April 1854), 181–2.

Notes on the Second Comet of 1854, *Monthly Notices of the Royal Astronomical Society*, **14** (9 June 1854), 222–4.

1856

Note on the Telescopic Appearance of the Planet Mars, *Monthly Notices of the Royal Astronomical Society*, **16** (9 May 1856), 188.

1857

Simple Method of Finding Focal Length of small Convex Lenses, *Monthly Notices of the Royal Astronomical Society*, **17** (10 July 1857), 269–70.

1858

An occultation of Regulus, *Monthly Notices of the Royal Astronomical Society*, **18** (11 June 1858), 289–90.

Observations of Comet V, 1858, *Monthly Notices of the Royal Astronomical Society*, **19** (12 November 1858), 21–4.

1859

Notices of Traces of Eruptive Action in the Moon, *Monthly Notices of the Royal Astronomical Society*, **19** (13 May 1859), 234–5.

Suggestions as to the Structure of the Tails of Comets, *Monthly Notices of the Royal Astronomical Society*, **19** (8 July 1859), 351–2.

1860

Note on the Minute Companions of σ Orionis, *Monthly Notices of the Royal Astronomical Society*, **20** (13 April 1860), 253.
On projection in lunar occultations, *Monthly Notices of the Royal Astronomical Society*, **20** (11 May 1860), 297.

1861

Note on one of the comites of β Geminorum, *Monthly Notices of the Royal Astronomical Society*, **21** (8 March 1861), 182.
On the form of the shadow of Saturn's ring, *Monthly Notices of the Royal Astronomical Society*, **21** (14 June 1861), 260–1.

1862

The Earth in the Comet's Tail, *Intellectual Observer*, **1** (February 1862), 63–7.
The Planets of the Month, *Intellectual Observer*, **1** (April 1862), 230–4.
Work for the Telescope – Planets of the Month – Double Stars, *Intellectual Observer*, **1** (June 1862), 373–5.
On the Great Comet of 1861, *Monthly Notices of the Royal Astronomical Society*, **22** (13 June 1862), 304–5.
Saturn's Ring – Double Stars – Occultations, *Intellectual Observer*, **1** (July 1862), 431–9.
Transit of the Shadow of Titan – Double Stars – The Moon – Occultations, *Intellectual Observer*, **2** (August 1862), 52–60.
Opposition of Mars – Double Stars – Occultations – The Comet, *Intellectual Observer*, **2** (September 1862), 131–40.
Double Stars – Occultations – The Earth in Opposition, *Intellectual Observer*, **2** (December 1862), 370–9.

1863

Uranus – Double Stars – Occultations, *Intellectual Observer*, **3** (February 1863), 51–9.
Uranus – Silvered Glass Specula – Occultations, *Intellectual Observer*, **3** (March 1863), 123–31.
Specula for Telescopes – Double Stars – Occultations – Transits of Jupiter's Satellites, *Intellectual Observer*, **3** (April 1863), 213–21.

Astronomical Notes. Planetary Systems among the Stars – Eclipses – Planets of the Month, *Intellectual Observer*, **3** (May 1863), 296–9.
Cometary Phenomena, *Intellectual Observer*, **3** (June 1863), 377–82.
Clusters of Stars and Nebulae – The Surface of the Moon, *Intellectual Observer*, **4** (August 1863), 56–64.
The Planet Mars: A Fragment, *Intellectual Observer*, **4** (October 1863), 182–93.

1864

Clusters and Nebulae – Occultations – The Achromatic Telescope, *Intellectual Observer*, **4** (January 1864), 448–57.
Clusters and Nebulae – Double Stars – Great Nebula in Orion – Comparison of Sun and Stars – Occultations, *Intellectual Observer*, **5** (February 1864), 54–62.
Clusters and Nebulae – Double Stars – Occultations, *Intellectual Observer*, **5** (March 1864), 138–40.
The Moon – Planets of the Month – Double Star – Occultations, *Intellectual Observer*, **5** (April 1864), 193–206.
Solar Observation – Transit of Jupiter's Satellites, *Intellectual Observer*, **5** (May 1864), 292–9.
On Certain Suspected Changes in the Lunar Surface, *Monthly Notices of the Royal Astronomical Society*, **24** (13 May 1864), 201–6.
Neighbourhood of the Lunar Spot Mare Crisium – Jupiter's Satellites – Occultations, *Intellectual Observer*, **5** (June 1864), 359–68.
Solar Observation – Colours of Stars – Constitution of Nebulae – Transits of Jupiter's Satellites, *Intellectual Observer*, **5** (July 1864), 434–43.
Double Stars – Colours of Stars, *Intellectual Observer*, **6** (August 1864), 58–61.
Colours of Stars – Clusters and Nebulae, *Intellectual Observer*, **6** (September 1864), 108–19.
The North-West Lunar Limb – Clusters and Nebulae – Occultations, *Intellectual Observer*, **6** (October 1864), 204–12.
The North-West Lunar Limb (continued) – The Achromatic Telescope – Occultation, *Intellectual Observer*, **6** (November 1864), 257–62.
Clusters and Nebulae – Double Stars – The Planet Mars – Occultations, *Intellectual Observer*, **6** (December 1864), 343–51.

1865

On a suspected Change of Brightness in the Lunar Spot, Werner, *Report of the Thirty-Fourth Meeting of the British Association for the Advancement of Science*, Bath, September 1864, **34** (1865), 8.

On the Invisible Part of the Moon's Surface, *Report of the Thirty-Fourth Meeting of the British Association for the Advancement of Science*, Bath, September 1864, **34** (1865), 9.

Herschel's Catalogue of Nebulae – The Achromatic Telescope, *Intellectual Observer*, **6** (January 1865), 446–55.

Lunar Arctic Region: Mare Frigoris: Mount Taurus: Posidonius – Occultations, *Intellectual Observer*, **7** (February 1865), 49–57.

Astronomical Notes, *Intellectual Observer*, **7** (March 1865), 133–42.

The Achromatic Telescope, Dialytes, and Fluid Lenses – Nebulae – Double Stars – Occultations, *Intellectual Observer*, **7** (April 1865), 179–90.

Lunar Taurus (S.) and Argaeus – Occultations, *Intellectual Observer*, **7** (May 1865), 255–8.

Dr. Draper's Telescope, *Intellectual Observer*, **7** (June 1865), 368–73.

Colours of Stars – Occultations, *Intellectual Observer*, **7** (July 1865), 467–71.

Lunar Details – Colours of Stars, *Intellectual Observer*, **8** (August 1865), 28–33.

Celestial Photography – Engelmann on Double Stars – Crimson Star, *Intellectual Observer*, **8** (September 1865), 133–43.

Clusters and Nebulae – Double Stars – Occultations, *Intellectual Observer*, **8** (October 1865), 207–16.

The Lunar Mare Serenitatis – Double Stars – Occultations, *Intellectual Observer*, **8** (November 1865), 292–301.

M. Chacornac on the Moon – Occultations, *Intellectual Observer*, **8** (December 1865), 370–5.

1866

Opposition of Ceres – Occultations, *Intellectual Observer*, **8** (January 1866), 454–60.

Lunar Details: Occultations, *Intellectual Observer*, **9** (February 1866), 56–64.

Notices of the Great Nebula in Orion, *Monthly Notices of the Royal Astronomical Society*, **26** (9 March 1866), 208–11.

Lunar Details, *Intellectual Observer*, **9** (April 1866), 173–81.

The Planet Saturn, *Intellectual Observer*, **9** (May 1866), 247–67; (June 1866), 366–81; (July 1866), 466–9.

The Planet Saturn, *Intellectual Observer*, **10** (August 1866), 49–59.

The Planet Saturn (continued), *Intellectual Observer*, **10** (September 1866), 142–9.

The Planet Saturn (continued), *Intellectual Observer*, **10** (October 1866), 194–202.

Cometary Light – Nebulae – Occultations, *Intellectual Observer*, **10** (November 1866), 281–8.
Nebular and Stellar Spectra – Solar Observations – Red Star – Planets – Occultations, *Intellectual Observer*, **10** (December 1866), 386–93.

1867

Lunar Details – Occultations, *Intellectual Observer*, **10** (January 1867), 441–3.
Light Spots in the Lunar Night – The Crater Linné – Occultations, *Intellectual Observer*, **11** (February 1867), 51–60.
Schröter's Meteors – The Lunar Cassini – Crimson Star – Occultations, *Intellectual Observer*, **11** (March 1867), 144–51.
Lunar Delineation – The Lunar Aristillus and Autolycus, *Intellectual Observer*, **11** (April 1867), 195–205.
Red Star – Double Stars – Nebulae – Linné and Aristoteles – Occultations, *Intellectual Observer*, **11** (May 1867), 275–83.
Lunar Appenines – Clusters and Nebulae – Occultations, *Intellectual Observer*, **11** (June 1867), 379–87.
Clusters and Nebulae – Southern Objects – Double Stars – Occultations, *Intellectual Observer*, **11** (July 1867), 459–67.
Mare Vaporum – Lunar Clefts – Occultations, *Intellectual Observer*, **12** (August 1867), 52–60.
The Lunar Clefts – Mare Vaporum – Jupiter's Satellites – Occultations, *Intellectual Observer*, **12** (September 1867), 95–104.
The Gruithuisen's City in the Moon – Jupiter's Satellites – Occultations, *Intellectual Observer*, **12** (October 1867), 214–23.
The Lunar Eratosthenes and Copernicus – Jupiter's Satellites – Occultations, *Intellectual Observer*, **12** (November 1867), 273–80.
Lunar Details – Double Stars – Clusters and Nebulae – Transits of Satellites – Occultations, *Intellectual Observer*, **12** (December 1867), 370–81.

1868

Lunar Sketches – Transits of Jupiter's Satellites – Occultations, *Intellectual Observer*, **12** (January 1868), 435–45.
Lunar Crater Linné, *Monthly Notices of the Royal Astronomical Society*, **28** (8 April 1868), 185–7; (8 May 1868), 218–9.
Globular Clusters of Stars, *The Student and Intellectual Observer*, **1** (July 1868), 454–61.
Beautiful Telescopic Field, *The Student and Intellectual Observer*, **2** (December 1868), 391.

1869

Linné – Hyginus – Diagonal Prism Eye Piece, *The Student and Intellectual Observer*, **2** (January 1869), 438–44.

Colour in the Moon, *The Student and Intellectual Observer*, **3** (May 1869), 251–60.

Work for the Telescope, *The Student and Intellectual Observer*, **3** (July 1869), 424–7.

American Photographs of Total Solar Eclipse of August 7, 1869, *Monthly Notices of the Royal Astronomical Society*, **30** (12 November 1869), 4–5.

Hints to Astronomical Students – Choice of Telescope, *English Mechanic*, **10** (17 December 1869), 327–8.

1870

Hints to Astronomical Students – Buying the Telescope, Quality, *English Mechanic*, **10** (21 January 1870), 448.

Hints to Astronomical Students – Centring the Lens, *English Mechanic*, **10** (25 February 1870), 569.

Hints to Astronomical Students, *English Mechanic*, **11** (10 June 1870), 74, 265.

Continuation of Last, *English Mechanic*, **11** (15 July 1870), 386.

Height of Clouds: Luminous Arches, *English Mechanic*, **11** (9 September 1870), 587.

Telescopic Objects, *English Mechanic*, **12** (4 November 1870), 154.

On the Planet Jupiter, 1869–70, *Popular Science Review*, **9** (1870), 127–37.

1871

Notice of a presumed new Variable Star in the Constellation of Orion, *Monthly Notices of the Royal Astronomical Society*, **31** (13 January, 1871), 84–6.

Solar Observations – Variable Stars, *English Mechanic*, **12** (17 March 1871), 606.

The Planet Jupiter, *Nature*, **3** (30 March 1871), 430–1.

Aurora by Daylight, *Nature*, **4** (11 May 1871), 27.

Notes on the Construction of Achromatic Object Glasses, *English Mechanic*, **13** (12 May 1871), 169–70.

The Berthon Dynamometer, *English Mechanic*, **13** (1 September 1871), 597.

Tests for the Telescope, *English Mechanic*, **13** (8 September 1871), 615.

A New Dynamometer, *Nature*, **4** (28 September 1871), 427–8.
The Light of Jupiter's Satellites, *Nature*, **4** (5 October 1871), 442–3.
The Berthon Dynamometer, *Nature*, **5** (2 November 1871), 6–7.
Astronomical Arrears, *English Mechanic*, **14** (22 December 1871), 354.
Star Colours, *The Student and Intellectual Observer*, **5** (1871), 481–91.
Observations on Jupiter in 1869–70, *Popular Science Review*, **10** (1871), 276–83.

1872

Work for the Telescope, *English Mechanic*, **14** (5 January 1872), 400.
Note on the Variable Star S Orionis, *Monthly Notices of the Royal Astronomical Society*, **32** (12 January 1872), 89–90.
Aurora Borealis of Feb. 4th, 1872, *Nature*, **5** (15 February 1872), 303.
The Meteor of March 4th, *Nature*, **5** (21 March 1872), 400.
Mädler's History of Astronomy, *Nature*, **6** (23 May 1872), 58–9.
Meteor, *Nature*, **6** (17 October 1872), 493.
The Swiss Sky – Stars in Corvus, *English Mechanic*, **16** (18 October 1872), 108.

1873

Star Shower in 1838, *Nature*, **7** (16 January 1873), 203.
Variable Star S Orionis – Ruby Star – Hollis's Astronomical Almanac, *English Mechanic*, **16** (7 February 1873), 506.
Earthquake in Pembrokeshire, *Nature*, **7** (13 February 1873), 283.
Astronomical Notices: Lunar Crater Moretus, *English Mechanic*, **17** (9 May 1873), 195.
'Lunar Nomenclature' denying responsibility for method of naming lunar spots, *English Mechanic*, **17** (19 August 1873), 607.
The Garnet Star in Cepheus, on alleged error in note in 2nd ed 'Cel Obs', *English Mechanic*, **18** (31 October 1873), 166.
How to Sketch the Moon, *Popular Science Review*, **12** (1873), 234–42.

1874

Mars, *Nature*, **9** (12 February 1874), 287–9.
Mars, part II, *Nature*, **9** (19 February 1874), 309–11.

1875

On Transit of Venus and Berthon Model, *English Mechanic*, **20** (1 January 1875), 397.

The Lunar Atlas, *English Mechanic*, **21** (19 March 1875), 11–12.
On the Star 61 Geminorum, *Monthly Notices of the Royal Astronomical Society*, **35** (9 April 1875), 340.
61 Geminorum, *English Mechanic*, **21** (7 May 1875), 196.
Meteors, *English Mechanic*, **22** (8 October 1875), 95.
Telescopic drawing of Orion Nebula and Cheap 3' Telescope, *English Mechanic*, **22** (19 November 1875), 255.
Lunar Mountains, *English Mechanic*, **22** (10 December 1875), 321.
Lunar Crater Cichus, *English Mechanic*, **22** (17 December 1875), 348.

1876

Meteor in the Daytime, *Nature*, **13** (6 January 1876), 187.
Note on the Variable Star S Orionis, *Monthly Notices of the Royal Astronomical Society*, **36** (14 January 1876), 107.
Caroline Herschel, *Nature*, **13** (9 March 1876), 361–3.
Astronomical Replies – Double Stars, Companion of Sirius, younger Tulley, eyepiece Webb owned, *English Mechanic*, **23** (17 March 1876), 14–15
Note on the Two Exterior Satellites of Uranus, *Monthly Notices of the Royal Astronomical Society*, **36** (12 April 1876), 294–6.
The Satellite of Venus, *Nature*, **14** (29 June 1876), 193–5.
Jupiter's satellites, *English Mechanic*, **23** (8 September 1876), 667.
Bright Meteor, *English Mechanic*, **24** (17 November 1876), 242.
Reports on Astronomy, *The Argonaut*, **3** (1876), 186–90, 314–16.

1877

Re Reflectors and Refractors, *English Mechanic*, **24** (19 January 1877), 455.
Zodiacal Light, *English Mechanic*, **24** (9 February 1877), 526.
Reports on Astronomy, *The Argonaut*, **5** (1877) 116 *et seq*.

1878

Astronomical Notes. Triangle in Monoceros; Doubles in Cassiopeia, *English Mechanic*, **26** (8 February 1878), 527.
Astronomical Notes. Answering Langdon 'The Dark Side of Venus', *English Mechanic*, **27** (15 March 1878), 16.
Dated 'Cheltenham' re Deterioration of Silvering of Mirror and Intrusion of Insects, *English Mechanic*, **27** (29 March 1878), 67.
Dark Side of Venus, *English Mechanic*, **27** (12 April 1878), 117.
The Organ at Uim, *English Mechanic*, **27** (26 April 1878), 169.

Double Star, *English Mechanic*, **27** (3 May 1878), 196.
Notes in Travelling, *English Mechanic*, **27** (19 July 1878), 469–70.
Meteor, *English Mechanic*, **27** (6 September 1878), 651.
Andromede Meteors, *English Mechanic*, **28** (6 December 1878), 320.

1879

Lunar Helicon, *English Mechanic*, **29** (4 April 1879), 87.
Celestial Objects, *English Mechanic*, **29** (15 August 1879), 568.
The Planets of the Season, *Nature*, **20** (23 October 1879), 605–6.
Reflectors and Refractors *English Mechanic*, **30** (7 November 1879), 211.
The Planets of the Season: Saturn, *Nature*, **21** (27 November 1879), 87–9.
Discovery of a Gaseous Nebula, *Nature*, **21** (4 December 1879), 111.
Discovery of a Gaseous Nebula, *English Mechanic*, **30** (5 December 1879), 308.
Discovery of a Gaseous Nebula in Cygnus, *Monthly Notices of the Royal Astronomical Society*, **40** (12 December 1879), 90–1.

1880

The Planets of the Season: Mars, *Nature*, **21** (1 January 1880), 212–3.
Gaseous Nebula, *English Mechanic*, **31** (19 March 1880), 34.
Celestial Objects for Common Telescopes, *English Mechanic*, **31** (26 March 1880), 66.
New Gaseous Nebula, *Astronomische Nachrichten*, **96** (1880), 191–2.

1881

FRAS re Delays to ‘Celestial Objects’, *English Mechanic*, **33** (24 June 1881), 380.
The Autumn Sky, part I, *Nature*, **25** (3 November 1881), 9–10.
The Autumn Sky, part II, *Nature*, **25** (10 November 1881), 36–9.
Gamma Andromedae, *English Mechanic*, **34** (2 December 1881), 301.

1882

Great Nebula in Andromeda, *Nature*, **25** (9 February 1882) 341–5.
Star Colours – Blue Star, *English Mechanic*, **34** (24 February 1882), 589.
The Planet Mars, 1881–2. Discussion of Schiparelli *et al.* on Observations of Canals, *English Mechanic*, **35** (31 March 1882), 79.
Canals on the Planet Mars, *Knowledge*, **1** (14 April 1882), 519.

Re Query 46647 14 Apr on _ Canari, R Leonis, a Double Star, *English Mechanic*, **35** (21 April 1882), 151.

Recent Discoveries in the Planet Mars, *Nature*, **26** (4 May 1882), 13.

Measurement of the focal length of deep convex lens, 'Including an account of a method of finding the focal length of a very small lens employed by me on Nov. 24 1835', *Knowledge*, **1** (18 May 1882), 605–6.

New Variable in Cygnis found by Birmingham about 14 months ago, not detected by TWW as too small or concealed by the trees which limit my telescope house, *English Mechanic*, **35** (28 July 1882), 474.

'Anomalous' Tail to the Comet 1882, *Nature*, **27** (23 November 1882), 89.

Great Comet. Copy of Letter to Times newspaper 12 Nov 1882, *English Mechanic*, **36** (24 November 1882), 272.

Earthquakes in Herefordshire, *Knowledge*, **2** (15 December 1882), 469–70.

The Great Comet, *Knowledge*, **2** (24 November 1882), 413; (22 December 1882), 480.

Mars, *Nature*, **27** (28 December 1882), 203–5.

1883

Optical Effects of Belladonna, *Knowledge*, **3** (2 February 1883), 68–9.

Astronomical – γ Virginis, *English Mechanic*, **37** (15 June 1883), 335.

Astronomical. Welcoming Amateur Measures of Faint Doubles and Multiples, *English Mechanic*, **38** (7 September 1883), 12.

On Lunar Delineation, *Knowledge*, **4** (16 November 1883), 302–3.

1884

Saturn's Ring, *English Mechanic*, **38** (29 February 1884), 557.

The Theory of Sunspots, *Nature*, **30** (15 May 1884), 59–60.

1885

Saturn, *Nature*, **31** (26 March 1885), 485–6.

Index

Abergavenny, 94, 99
aberrations, 123, 125
Adams, John Couch, 104, 116
Airy, George Biddell, 19, 131
Allen, David, 237, 238
antiquities, 32, 74, 75, 76, 77
Argelander, W., 208
Ashmolean Museum, 19
asteroids, 151
aurorae, 148
Aveline, Fanny, 25
Aylmer, Geoffrey, 80, 82

Baily, Francis, 207
Ball, Robert S., 216, 222, 225
Bardou, 131
Barker, Edmund, 238
Barlow, Peter, 124, 126
Baron, John, 25
Bate, Robert Brettell, 122, 125, 126, 141, 170
Baxendell, J., 147
Beaufort, Duke of, 30
Bedford Catalogue, 129, 146, 200, 209, 219
Beer, W., 153, 156, 158, 159, 160, 161
Benitez, Merchan, 213
Berthon, Edward Lyon, 37, 38, 111, 112, 119, 128, 134, 139, 143
Bevan, Louise, 52
Biela, Wilhelm von, 182, 183
Biela's comet, 183
Birch, John, 78, 79, 80
Birmingham, John, 210
Birt, William Ratcliff, 159, 160, 162, 163
Bishop of Hereford, 44, 60, 70, 71
Bishop of Worcester, 108
Bishop, George, 140
Bishop, J., 9, 10, 11
Black Mountains, 29 36
Blair, Robert, 123, 124
Blunt, J.J., 71
Bodleian Library, 4, 13, 77
Bond, George P., 183, 228
Bonnor, Charles, 10, 11
Brampton Bryan castle, 81, 82
Breautè, Nell de, 182
Bredichin, Theodor, 109
Breen, James, 133
Brewer, Ebenezer, 221
British Association for the Advancement of Science, 19, 28, 32, 157, 159, 162, 207, 215
British Astronomical Association, 40, 107, 109, 110, 128, 131, 138, 143, 192, 235, 236, 237

British Library, 97
British Museum, 77
Brorsen's comet, 185
Brougham, Henry, 206
Brown, Elizabeth, 109
Browning, John, 137, 138, 141
Buckland, William, 17
Buczynski, D.G., 143
Burnham, Sherburne W., 201, 207, 208, 209
Burton, Charles, 175, 176
Burton, Edward, 60
Butler, Joseph, 12

Callan, Nicholas, 104, 116
Calver, George, 53, 113, 135, 137, 138, 139, 201
Camden Society, 4, 78, 79
Cardiff, 4, 41, 63
Catholic Emancipation Act, 16, 65, 103
Ceres, 151
Challis, James, 133
Chambers, George F., 37, 40, 144, 220
Cheltenham, 50, 51, 91, 94, 96, 149, 217
Cheltenham Ladies' College, 28, 41, 63, 217
Cheltenham Working Man's Club, 28
Chevallier, Temple, 103
Clapham School, 108
Clark, Alvan, 110, 131, 132, 133, 136, 140, 142, 185, 207
Clerke, Agnes, 109, 165, 216, 222, 223
Clyro, 42, 53, 68, 107
Coggia's comet, 104
comet 1849 G1, 183
comet Bennett, 182
comet Hale–Bopp, 181, 187, 190
comet Hyakutake, 181, 187, 188
comet Ikeya–Zhang, 181
comet Machholz, 181
comet Swift–Tuttle, 191, 192
comet Thatcher, 185
comet VI 1863, 192
comets, 11, 12, 28, 33, 39, 104, 132, 133, 149, 150, 171, 173, 181, 217, 218, 226
see also Great Comets
constellations, 204, 205
Cortie, Aloysius, 109
Cox, David, 31
Crabtree, William, 101
Crichton, Mrs, 48
Cromwell, Oliver, 2
Crossley, Edward, 162
Crowe, Michael, 176

Dall, Horace E., 131, 143
Dalton, John, 19
D'Arrest's comet, 185
Darwin, Charles, 59, 64, 156, 221
Dawes, William Rutter, 73, 102, 110, 115, 132, 133, 140, 142, 165, 166, 179, 198, 201, 204, 207
De La Rue, Warren, 133, 140, 162, 165
De La Touche, Mr, 148
De Vico, Francesco, 167
Dearborn Observatory, 209
Dembowski, H., 207
Dew, Fanny, 52, 53
Dewhirst, David W., 145
dialytes, 126
Dick, Thomas, 105
Dimlands Castle, 94
discoveries, 211
Dollfus, Audouin, 169
Dollond, George, 124, 126, 140
Dollond, John, 123
Dollond, Peter, 127, 130
Donati's comet, 132, 133, 149, 150, 185, 186, 189
Doolittle, E., 210
double stars, 147, 200, 217, 238, 239, 240
Downside Abbey, 104
Dreyer, J.L.E., 175
Duff, Helen, 25, 26, 30, 61

earthquakes, 51, 85, 148
eclipses, 109, 149, 194, 198, 199

Einstein, Albert, 199
electrotherapy, 28
Ellis, W., 140
Encke's comet, 192
Espin, Thomas H.E.C., 32, 46, 47, 52, 53, 65, 67, 104, 106, 107, 113, 114, 139, 143, 145, 148, 151, 170, 209, 220

Faraday, Michael, 19, 20, 136
Flamsteed, John, 205
fluid lenses, 124, 125, 126
Foucault, J.B.L., 134
Foulkes, T.H., xvii, 40
Fouracres, Mr, 125
Frankland, Edward, 136, 137
Franks, William S., 209

Gadbury, John, 173
Ganarew, 32, 33, 63, 75
gardening, 49
Gascoigne, William, 101
Genner, Ian, 237
geology, 153, 154, 221
George III, King, 2
George IV, King, 205
Gill, David, 134
Gilly, Dr, 103
Gladstone, William, 107, 118
Gloucester, 9, 32, 44, 61, 156
Gloucester Cathedral, 4, 31
Gloucester Literary and Scientific Association, 28
Glyn Jones, Kenneth, 235, 236, 237
Golden Valley Railway, 29
Graham, Mr, 183
Great Comet of 1811, 181
Great Comet of 1819, 182
Great Comet of 1823, 182
Great Comet of 1854, 184
Great Comet of 1860, 185
Great Comet of 1861, 39, 187, 192
Great Comet of 1882, 149, 192, 193
see also comets
Great June Comet (1845), 182, 183
Green, Nathaniel E., 138, 175
Gregory, David, 123
Grubb, Thomas, 135, 141
Guy's Hospital, 33

Haileybury College, 104
Hall, Chester Moor, 123
Hall, Frederick, 104
Halley's comet, 181, 187
Harding, Sarah (see Webb, Sarah)
Harewood, 27
Harvard College Observatory, 132
Harvey, Susan, 23, 25
Hawkins, Hester Periam, 107
Hawkins, Joshua, 107
Hay-on-Wye, 36, 41, 96
Henslow, George, 221
Hereford, 36, 60, 61, 75, 98, 134
Hereford Cathedral, 26, 45, 69, 71, 79, 111, 138
Hereford Literary and Philosophical Society, 41, 56
Herschel, John F.W., 108, 122, 129, 142, 159, 162, 166, 195, 197, 239
Herschel, William, 121, 152, 176, 177, 194
Hertford College, 12
Hill, Thomas, 8
Hind, John Russell, 140
Hockey, Thomas, 176
Holder, Miss J., 23, 25
holidays, 49, 50
Horrocks, Jeremiah, 101
Hoskyns, Hungerford, 27
Howlett, Jem, 11
Huggins, William, 135, 140, 141, 222
Humboldt, Alexander, 170
Hutchinson, Neville, 221
Hutton, Ronald, 82
Huxley, Thomas H., 136, 225, 230
Hynes, Stephen, 238, 239
Hysom, E.J., 237

illness, 51
International Astronomical Union, 236
Inwards, Richard, 131
Isles, John, 237, 238

James, William, 16
Jamieson, Alexander, 204, 205, 206
Jenner, Edward, 25, 34
Johns, Charles A., 221
Jones, John, 105
journals, 217, 218
Jupiter, 148, 149, 151, 166, 167, 176, 177, 217, 225, 227

Kay, Jenni, 239
Kepler, Johannes, 173
Key, Henry Cooper, 46, 111, 134, 137, 138
Kilvert, Francis, 29, 42, 44, 45, 47, 50, 68, 75, 107
King's College, London, 103
Kingsley, Charles, 221
Kirchhoff, Gustav, 141
Kitchiner, William, 130
Knott, George, 147, 207, 209
Kronk, Gary, 184

Langdon, Roger, 106
Langley, S.P., 165
Larard, John, 236, 237
Lassell, William, 104, 116, 122, 133
Leamington, 29
lectures, 217
Lee, John, 146, 207
lenses, 123
Leominster, 90, 96, 97
Lick Observatory, 201
Liebig, Justus von, 134
Little Doward, 75
Liverpool Astronomical Society, 47, 109
Llanthony Abbey, 29
Lockyer, Norman J., 217, 218
Lowell, Percival, 169

M13, 111, 223, 230
Maclear, Thomas, 130
Mädler, J.H., 153, 156, 158, 159, 160, 161, 162, 171, 172
Magdalen Hall, 12, 13
Manchester Astronomical Society, 109, 110
Mariner 2, 169
Markree, 183, 195
Married Women's Property Act, 51, 68
Mars, 173, 174, 177, 179, 218, 222, 225, 227
Matthews, Revd, 135
Mauvais, F.V., 183
Maynooth Seminary, 104
McBride, John, 12, 13, 19
Mee, Arthur, 36, 37, 38, 40
Melpomene, 151
Mercury, 32, 149, 167, 170, 171, 194, 198, 199
Merz, 132
Messier objects, 235
meteorology, 5, 7, 85, 86, 148, 159, 173
meteors, 2, 28, 37, 122, 166, 183, 187
Milburn, William, 145
mirrors, 122, 125, 133
Mitchel Troy, 30, 31, 32, 52, 53
Moon, 2, 10, 28, 33, 122, 133, 147, 151, 152, 166, 167, 215, 216, 217, 218, 226
Moon Committee, 162, 163
Moore, E.G., xviii, 238
Moss, Mr, 10, 11
Muirden, James, 236, 237

Nasmyth, James, 122
Neate, Alfred N., xviii, 57
nebulae, 141, 147, 151, 166, 171, 200, 211, 217, 228, 229, 236
Neison, E., 57
Neptune, 141, 151, 179
Newkomm, Chevalier, 4
Newman, John Henry, 104
Newton, Isaac, 121, 123
NGC 7027, 211
Nichol, John P., 230
Noble, William, 195
Nova Lacertae, 113
Nova Ophiuchi, 140

Oates, Colonel, 76
observatory, 38, 111, 112, 128, 139, 144

optics, 123, 218, 226
Orcop, 74
ordination, 61
organs, 29, 32, 65, 70
Orion Nebula, 141, 166, 228, 229
Ouseley, Frederick Gore, 73
Oxford, 12, 60
Oxford University Observatory, 108

Pallas, 151
Palmer, Charles S., 52
Pearson, William, 107, 108, 110
Peel, Robert, 16
Pencoed, 26, 27, 60, 67, 68
Penoyre, Anna Maria, 63
Penoyre, Frances R.B.S., 33, 36, 41, 46
Perry, J.J. Muschamp, 138, 229
Perry, Stephen J., 103, 109, 118
Petersen, A.C., 184
Phillips, John, 153, 159, 162, 166
Phillips, Theodore E.R., 114
photography, 48, 140, 223
Piazzi, Guiseppe, 209
planets, 10, 133, 140, 141, 148, 149, 151, 153, 165, 179, 217, 218, 222, 225, 226, 227, 228
Plössl, G.S., 129
Pogson, Norman R., 140
Pons, Jean Louis, 182, 183
Powells, Marie, 65, 67
Pretty Polly, 7
Pritchard, Charles, 108, 110
Proctor, Richard A., 158, 204, 207, 216, 218, 220, 222, 223, 224
Purchas, W.H., 184, 185

Queen's University, Belfast, 103

Radcliffe Observatory, 13, 17, 18, 125
Raffles, Thomas, 115
railways, 29
Ranters, 65, 67, 72
Ranyard, Arthur C., 4, 29, 32, 37, 38, 39, 40, 41, 44, 46, 47, 48, 49, 51, 52, 63, 64, 65, 68, 71, 78, 79, 80, 82, 132, 133, 136, 138, 139, 217, 230
red stars, 210
Rigaud, Stephen, 18, 20, 125
Robertson, John, 105
Robinson, Thomas Romney, 103, 110
Roe, John, 78
Rogers, A., 126
Romsey Abbey, 111, 119
Rosse, Lord, 111, 121, 122, 134, 162, 239
Ross–Gloucester railway, 29
Ross-on-Wye, 1, 2, 3, 8, 75
Royal Astronomical Society, 33, 41, 101, 107, 108, 110, 113, 114, 127, 130, 131, 132, 135, 138, 145, 149, 154, 167, 198, 206, 235
Royal Military Academy, 124
Royal Observatory, Greenwich, 18, 32, 238
Royal Society, 103, 108, 124, 135, 141, 159
Royal Society of Chemistry, 136
Rudwick, Martin J.S., 154
Ruskin, John, 152
Rutherfurd, Lewis, 162

S Orionis, 147, 203, 211, 213
Sabine, Edward, 159
Sadler, Herbert, 46, 57, 147
Saturn, 133, 140, 151, 167, 177, 178, 179, 227, 228
Schiaparelli, Giovanni V., 175
Schjellerup, H.C.F.C., 210
Schröter, Johann H., 152, 153, 157, 160, 161, 162, 165, 166, 167
Schumacher, H.C., 130
Schweizer, K.G., 183
Science Museum, 131
Scott, Mrs Acton, 80
Sedgwick, Adam, 156
Selenographical Society, 57
Sheldonian Theatre, 15, 16, 20
Sidgreaves, Walter, 109
Sise, Anne (grandmother), 2
Sise, James (great-grandfather), 2
Smith, William, 20
Smyth, William Henry, 129, 130, 146, 147, 200, 207, 213, 219

Society for the Diffusion of Useful Knowledge, 206
Society of Antiquaries, 4, 80
Society of Ordained Scientists, 114
spectroscopy, 141, 222, 223
speculum metal, 121, 133, 134
St Paul's School, 3
St Weonards church, 27, 29, 30, 70, 76
star clusters, 200, 217
star colours, 213
Steinheil, C.A. von, 134
Stokesay, 90, 97, 148
Stonyhurst College, 103, 104, 108, 109, 110, 133
Stretton Sugwas, 46, 111, 134
Stroobant, Paul, 169
Struve, F.G.W., 204
Struve, Otto, 213
Sun, 10, 149, 157, 167, 194, 217, 218, 226, 227, 228
Sutherland, J., 145
Symonds, Mr, 62

Tebbutt, John, 187
telescopes, 11, 32, 37, 121, 201, 217, 218
Temple, Frederick, 106
Tenby, 32, 49
Thompson, Flora, 106
Three Choirs Festival, 10
Tomes, Robert, 13, 14, 16
Tow Law, 113, 145
Towneley, Richard, 101
transit of Mercury, 32, 149, 194, 198
Tretire, 2, 3, 7, 8, 12, 26, 31, 32, 33, 36, 41, 63, 68
Trinity College, Dublin, 103, 136
Trollope, Anthony, 102, 106
Troy House, 30, 31, 32, 42
Tudor, Owen, 30
Tudor, Thomas, 30, 36
Tulley, Thomas/William, 32, 65, 127, 129, 131, 147, 200
Tuttle's comet, 185
Tyndall, John, 136
Tyn-y-beddau, 75

Underwood, Thomas, 1, 3
United States Naval Observatory, 240
universities, 103, 104, 105
University College London, 103
University College, Durham, 103
University of Edinburgh, 124
Uranus, 151, 177, 217

Vargas, Jurado, 213
variable stars, 147, 210, 238
Vatican Observatory, 114, 119, 120
Velikovsky, Immanuel, 171
Venables, Richard L., 53, 107
Venus, 10, 149 166, 167, 168, 169, 170, 171, 172, 173, 199
Vesta, 151
Vienna Observatory, 130

Wadham College, 3, 12
Walker, Tim, 239
Wall, John, 129
Waller, C., 143
Waterton, Charles, 115
Webb, Anne Frances (sister), 3
Webb, Henrietta Montagu (née Wyatt, wife), 30, 31, 32, 33, 36, 40, 41, 45, 46, 47, 48, 49, 50, 51, 52, 81, 82, 96, 131, 140, 147, 149, 187, 192
Webb, Henry (of Odstock), 20
Webb, John (father), 2, 3, 4, 8, 9, 10, 12, 15, 18, 25, 26, 27, 30, 31, 32, 33, 36, 37, 41, 42, 60, 62, 63, 70, 72, 74, 75, 77, 78, 79, 81, 82, 126, 129, 139, 181
Webb, Sarah (née Harding, mother), 3, 9, 10, 15, 16, 19, 20, 25, 26, 28, 31, 32
Webb, Thomas (great-uncle), 2
Webb, William (grandfather), 2
White, Gilbert, 165
White, Mr, 10, 11
Willis, James, 36, 49
Willis, Robert, 19
Wilson effect, 196
With, George Henry, 37, 46, 52, 100,

111, 128, 134, 135, 136, 137, 138, 139, 201
Wollaston, Francis, 130
Wollaston, William Hyde, 130
Wood, John G., 221, 222
Woolhope Club, 75
Wyatt, Arabella, 30
Wyatt, Arthur, 30, 31
Wyatt, Emma, 51
Wyatt, Helen (Helen of Troy), 42, 53
Wyatt, Henrietta Montagu (see Webb, Henrietta Montagu)
Wyatt, Matthew Digby, 31
Wyatt, Thomas Henry, 31, 33, 41, 56
Wyatts of Mitcheltroy, 30, 31, 42, 43, 45, 51
Wyesham, 30

zodiacal light, 33, 148